FORAGING

Recognizing Toxic and
Poisonous Wild Plants and Mushrooms

MONA GREENY

Table of Contents

Introduction

If you want to find which plants are poisonous or which mushrooms can kill you, read about it in this book. Here in this book, "Poisonous Plants and Mushrooms," you have a ready reference to the common poisonous plants and dangerous mushrooms.

This is a detailed guide that helps you learn to recognize those plants that can give you a rash or disrupt your digestion. Learn about the toxins that reside in the plants and mushrooms near your house. Once you learn about the common toxins and the method of dealing with them, you learn to avoid those that do you harm.

Learn about the importance of cooking your dishes well. Some toxins get destroyed by heating and some don't. Read about it all in this book.

There are some sample recipes also to help you know how to handle mushrooms and prepare delicacies at home, adding more variety and taste to your diet.

Begin your journey into the wonderful world of plants and mushrooms and gain the knowledge you need to deal with the deadlier kind. It pays to remain safe while exploring new cuisine and use exotic mushrooms in your kitchen.

Take your first step toward your new adventure with mushrooms and plants. I hope you have an exciting and wonderful journey!

Chapter One

Poisonous Plants

Nerium Oleander (Dogbane)

Popular in parks, avenues, and gardens, Nerium Oleander is an evergreen shrub used for ornamental purposes and is part of the Dogbane (Apocynaceae) family. It sports brilliant yellow, red, pink, or white clusters of flowers that grow at the end of their branches. Popular for its hue and fragrance, they are also a serious source of plant poison. All its parts are toxic - flowers, leaves, stem, and roots.

Description

The shrub grows up to 20-25 feet in height. It flowers in summer in white, pink, and red. The diameter of the 5-petalled flowers is 2.5-5 cm with 5-lobed deep corolla around the central tube. It might have a sweet scent. The fruit is long and narrow; the follicles are 5-22 cm. It splits open when mature to release downy seeds.

The thick leathery leaves are opposite in pairs or whorls of three. They are narrow, lancelot 5-20 cm long, and 1-3.4 cm broad. You find an entire margin having minute reticulate venation. Its stems are erect with glaucous bloom in the first year.

Habitat and Distribution

Native to India, China, Morocco, Portugal, and Mauritania, you can also find sporadic appearances of the Nerium Oleander in the Sahara. In the U.S., it appears on Virginia Beach in the north while you can find them planted along the roadsides and highways of California in the west. There are median strips present in Texas in the south planted as an aftermath of the 1900 hurricane.

Toxicity

Upon consuming any part of this plant, the person exhibits cardiac abnormalities or gastrointestinal symptoms. This is due to the presence of cardiac glycosides. Bradydysrhythmias or Tachydysrhythmias alongside a heart block and slowed conduction occurs. When there is severe poisoning, hyperkalemia could happen. The cardiotoxins of Oleander get absorbed transcutaneously as well as through inhalation routes.

Antidote

One common antidote for this poison is digoxin specific Fab fragment.

Medicinal Preparations

Used in native medicine for treating parasitic infestations, myalgia, cardiac diseases, as an abortifacient, and wound healing, among others.

Uses

People use this plant for screening their garden space, and you can grow it without protection. But it will die in severe winter conditions.

The Suicide Tree - Cerbera odollam (Apocynaceae)

The Suicide Tree or othalanga belongs to the Apocynaceae family. This infamous tree grows on the western coast of India, Kerala and its softball-sized fruit can cause the heart to stop. Many people have used this fruit to commit suicide. This dicotyledonous angiosperm is the perfect murder weapon, according to many people.

Description

This plant can grow up to 30 meters in height. Branchlets of the Cerbera Odollam remain whorled around the trunk. It has terminally crowded leaves with tapered bases, while the leaves have entire margins with acuminate apices. The plant produces a white, milky latex. Its fruit, when it is green, looks like a small mango.

With its fibrous shell, it encloses the ovoid kernel 2x1.5 cm that has two cross-matching halves.

Habitat and Distribution

It is a native of India and regions of southern Asia. Cerbera odollam prefers salty, coastal swamps and low-lying marshy areas. We can see it in Vietnam, Myanmar, the Philippines, Indonesia, Malaysia, and Thailand.

Toxicity

Inside the kernels of Cerbera odollam, we find a digoxin type of cardenolide called cerberin and cardiac glycosides. They disrupt the heartbeat by blocking the calcium ion channels in the heart.

Symptoms

Common symptoms of toxicity include vomiting along with thrombocytopenia. The fatal dose is in one kernel, and death occurs within 1-2 days. Here is the list of symptoms:

- Violent vomiting.

- A burning sensation inside the mouth.

- Coma leading to eventual death.

- Irregular heartbeat.

- Headache.

- Irregular respiration.

Treatment

The use of digoxin immune fab helps in the management of cerberin toxicity. We also need to address the hyperkalemia and bradycardia issues.

Uses

Substances found in the tree and its parts help produce deodorants, bioinsecticides, and rat poison. People grow it as a hedge plant between house compounds. They make purgative and emetic medicinal preparations from the latex, leaves, and bark.

Castor Oil Plant - Ricinus communis (Euphorbiaceae - Spurge)

The castor oil plant has features like oleander, and due to its seeds, it holds the world record for being the most poisonous plant. Ricinus Communis belongs to the Euphorbiaceae or spurge family.

Toxicity

Inside the seeds, we see very rich toxin ricin. The lethal dose is 4-8 seeds. Death will result if one doesn't act after eating the seeds. One will experience bloody diarrhea within 30-36 hours, along with intense abdominal pain and burning sensation within the mouth. This leads to death within 3-4 days.

Description

This suckering shrub is fast-growing, reaching heights of 38-42 feet. Even though it grows to almost becoming a tree, it cannot survive the severe cold. It has glossy leaves 17-42 cm in length that have 5-11 deep lobes. The leaves are palmate, alternate, and long-stalked. In some varieties, young leaves are bronze or reddish-purple and then go on to becoming dark green sometimes with a red tinge.

The Appearance of the Foliage

It is not uncommon to find dark-leaved plants alongside those with green leaves. Some have showy fruit capsules compared to the flowers. Flowers are unisexual but without petals and both male and female flowers occur on the same plant. The panicle-like inflorescence of the male flower is yellowish-green, numerous, and has prominent creamy stamens. In the female flower that lies within immature capsules with spines, we see prominent red stigmas.

The fruit is greenish to reddish-purple, shiny, oval, and large. It is common to remove these because ricin gets concentrated on the spines. The very poisonous seeds display a caruncle, which is a warty appendage useful for the dispersion of the seeds.

Medicinal Use

In low doses, Ricinus can affect the central nervous system, while the leaves show anti-microbial properties. It is useful as a decorative plant in public parks since Edwardian times in Ontario and Toronto in Canada. It grows wild in Southern California.

Habitat and Distribution

The main regions in the world that cultivate the castor plant include India, Brazil, and China. But the plant grows natively in the southeastern Mediterranean Basin, and eastern Africa.

Beach Apple - Hippomane mancinella (Euphorbia)

Sometimes referred to as the 'little apple of death', beach apple or poison guava, Hippomane mancinella of the Euphorbia family, helps Carib Indians poison the tips of their arrows. This Manchineel tree is so poisonous that merely brushing up against the bark will bring a violent allergic reaction. The skin in contact with the water that ran off from the tree will become blistered.

Description

This evergreen tree will grow up to 50 feet in height. It has shiny green leaves and a reddish-gray bark. The flowers are small and greenish-yellow. The fruits are apple-like to look at and are green or greenish-yellow when they ripen.

Toxicity

In the beach apple, phorbol is the powerful irri tant that causes allergic reactions in the skin. It is one of the many toxins found in the milky sap of the tree. This sap is present everywhere in the tree - the bark, fruit, and leaves. It becomes fatal due to the physostigmine content that creates a variety of reactions such as these:

- Diarrhea

- Vomiting

- Seizures

- Nausea

It could lead to death, so don't mess around with it. The poison is so strong that it can damage the paint of a car parked beneath the tree.

The fruit is pleasantly sweet in the beginning and then shows a strange peppery feeling. It can produce shock with bacterial superinfection. Burning the tree will cause acute keratoconjunctivitis if the smoke reaches the eyes. If you continue to eat the fruit, one reaches the stage where a huge lump in the throat and excruciating pain will stop the person from swallowing food.

Uses

They serve as natural windbreaks and for retarding beach erosion. Its roots help stabilize the loose soil. When used for making furniture, they leave the wood out in the sun to dry to remove the sap completely. Gum made from the bark helps treat edema while they use the dried fruits as a diuretic.

Habitat and Distribution

It is native to Central America and the tropical parts of southern North America. You also find it in northern South America, Florida, the Bahamas, Mexico, and the Caribbean. It grows well in brackish swamps and coastal beaches.

Poopy - Chelidonium majus (Papaveraceae)

This herbaceous perennial belongs to the poppy (Papaveraceae) family. Everything in the plant is toxic. People have used this plant since the days of Pliny the Elder during the first century. Other names for this plant are tetterwort, swallowwort, nipplewort, and greater celandine.

Description

This is a perennial herb that grows up straight and reaches 33-123 cm in height. It has blue-green pinnate leaves with wavy margins and lobes. The sap is yellow to orange in color. The flower has four yellow petals 18 mm long. We get flowers in late spring up to summer and the umbelliform cymes have four flowers. The seeds are small and black enclosed in a long capsule in the shape of a cylinder.

Habitat

This plant grows natively in most parts of Europe. You can also see it in parts of Africa, such as Morocco, Algeria, and Micronesia. In western Asia, we see it growing in Turkey, Iran, Siberia, Mongolia, Georgia, and Armenia, among other places.

Toxicity

The entire plant is toxic due to the presence of isoquinoline alkaloids. The main alkaloid is coptisine.

Use

Treats gallstones and dyspepsia and gets rid of warts. Treats various kinds of inflammatory disorders, including atopic dermatitis. People also used it as a foot refresher.

Rosary Pea - Abrus precatorius (Fabaceae)

One must have seen the brightly colored black, red, and white beads of the Abrus precatorius on many percussion instruments. The seeds are so poisonous that eating even one seed proves fatal. It also goes by the name of Rosary Pea, Crab's eye, jequirity, love pea, coral bead, John Crow bead, Indian licorice, gidee gidee, and Jamaica licorice. It belongs to the bean family, Fabaceae.

Description

This poisonous vine is broadleaf, deciduous, and evergreen. It forms deep roots and becomes invasive as birds spread it. It has flowers in pink, white, purple, and lavender colors. These remain clustered in the axils of the leaves. The leaves are palmately, pinnately, and bipinnately. The leaf arrangement is opposite, and the shape is oblong-ovate. Leaf margins are entire, and leaf length is 7.6 to 20 cm. Leaves are compound with 5-15 pairs of leaflets having a length of less than 1 inch.

Fruit is black, red, and burgundy and 1-3 inches long.

Habitat and Distribution

This plant is native to Australia and Asia. It grows in the wild in Jamaica, Java, and India.

Toxicity

Abrus precatorius has high severity toxicity characteristics with the high abrin content making the seeds toxic. This chemical is like ricin present in castor seeds. Abrin stops the action of ribosomes in the body; these ribosomes help you synthesize proteins. One seed is enough to cause a lethal effect as abrin is 75 times more toxic than ricin.

Uses

They use the seeds to make rosary necklaces. Since they are consistent in weight, the seeds help to weigh gold in a measure called Ratti. Eight Ratti gives you one Masha, and 12 Mashas make one tola. People of the Chamar caste of India use it for poisoning cattle and taking their hides. The white variety of this seed helps make an aphrodisiac oil.

Dumbcane - Dieffenbachia amoena (Araceae)

A strong species that can survive in any light conditions, Dieffenbachia or Dumbcane, is a houseplant that can grow as tall as 6 feet. In some parts of the world, people call it mother-in-law's tongue. Though it is poisonous, it rarely kills humans. It belongs to the Araceae family. They are notorious as being poisonous to dogs and cats.

Description

This perennial, herbaceous plant has a straight stem with leaves that are simple and alternate. They contain flecks and white spots, adding to their attractive appearance. The leaves have yellow coloring near the vein and dark green near the edges.

The random transition in each leaf creates ribbons and spots of yellow and green color along the body of the leaf. The leaf will grow to be 20 cm and 25 cm wide. The plant will rarely flower, and the blooms are green and unimpressive. It will grow in any soil condition.

Habitat and Distribution

We find it growing natively in many Caribbean Islands and Brazil. This includes Puerto Rico and tropical parts of New Mexico.

Toxicity

There is enough toxicity in the plant to kill a child in one minute and an adult in 15 minutes. It is because of calcium oxalate that forms razor-sharp crystals called raphides. It causes the tongue and mouth to swell and choke you.

Use

People use it as a houseplant. This shade-loving house plant adds pep to the interiors.

Angel's Trumpet - Brugmansia (Solanaceae)

People consider these flowers will herald one to their afterlives and so it also has the name Angel's Trumpet. The flowers are trumpet-shaped and hang down, but there is no doubt that the plant is poisonous. They belong to the nightshade (Solanaceae) family.

Description

This group of shrubs and small trees with semi-woody and branched trunks. They are less than 26 feet in height with toothed or entire simple leaves. The leaves that alternate are 10-32 cm long and 4-17 cm wide. The margin is entire or coarsely toothed with plenty of fine hairs. The flowers are pendulous with a fused corolla shaped like a trumpet. These are 14-45 cm long and 10-35 cm at the opening. The colors range from yellow, cream, white, green, pink, red, and orange. The flowers might be single or double.

Toxicity

Everything in Angel's Trumpet plant is poisonous. The toxicity is due to alkaloids hyoscyamine, scopolamine, and atropine. Upon eating a part of the plant, the person experiences disturbing hallucinations, memory loss, tachycardia, and paralysis.

Habitat and Distribution

Brugmansia grows natively in South America from northern Chile to Colombia and you can also find them along the Andes in Venezuela and south-eastern Brazil. They prefer sloping terrain with warm, humid days and cool nights.

Use

People use the flowering plants as ornamentals in frost-free climates and greenhouses. The alkaloids have medicinal value as an antiasthmatic, spasmolytic, anesthetic, narcotic, and anticholinergic agents.

White Snakeroot - Ageratina altissima (Asteraceae)

Poison from this plant comes through a third party, usually a cow. The milk of the cow that has ingested Ageratina altissima develops shakes. And, if the cow is lactating, the poison enters the milk. If anyone is unfortunate enough to drink that milk, he or she will develop sickness and could even die. This plant belongs to the Asteraceae family. Other names for it include White Snakeroot, white sanicle, and richweed.

Description

The plant is upright and can grow up to 1.5 meters in height. They blossom in late summer or fall. Flowers are white and appear bright. Seeds get propagated when fluffy groups get released into the breeze. The loose, flattened clusters of white flowers are 3-4 inches supported on stems that are 3-5 feet tall. Its dark green leaves are long-stalked with sharp teeth. The shape is like a lance or elliptical oval and taper pointed.

Habitat and Distribution

This poisonous perennial herb is native to eastern and central North America. They grow in brush thickets and woods and are adaptable to different growing conditions. It will also grow in open ground areas with a little shade. It is common throughout Missouri in the U.S. these days.

Toxicity

Tremetol is the toxin present in White Snakeroot. Abraham Lincoln's mother died of milk sickness caused by this plant. Ingesting this poison can lead to terrible intestinal problems, vomiting, and trembling.

The plant is poisonous to cattle, goats, sheep, and horses. Signs of poisoning include nasal discharge, wrong placement of hind feet (either too close together or too far apart), rapid or difficult breathing, arched body posture, and excessive salivation.

Use

People grow it in cottage gardens and wild gardens to provide a border for their garden. Root tea helps treat fever, kidney stones, and diarrhea. There is a popular misconception that one can use this to treat snakebites.

Marsh Marigold - Caltha palustris (Ranunculaceae)

You can see these lovely flowering plants beside streams and ponds, as they like oxygen-rich water and rich soil. The common names are marsh marigold and kingcup. It belongs to the Ranunculaceae family.

Description

The hairless perennial plant grows 12-77 cm tall. Its alternate true leaves form a rosette; the stems are hollow and erect. The branching roots are 2-3 mm thick. The leaf blade is 3-25 cm long and 3-20 cm broad. It has a blunt tip and a heart-shaped foot. The young leaves remain protected by a membranous sheath that could grow up to 3 cm in a grown plant.

Flowers are 2-5.4 cm in size, with 5-8 petals and they have bright colors like white, yolk yellow, or magenta. Real petals are lacking. There are 50-250 stamens.

Habitat and Distribution

They inhabit the temperate regions of the Northern Hemisphere. You can find it in the Himalayas, Bhutan, and Pakistan.

Toxicity

Among the many toxic compounds, the most significant one is protoanemonin. Ingesting large amounts of this plant will cause convulsions, vomiting, fainting, dizziness, and bloody diarrhea.

Use

Early spring buds and greens are edible when cooked. It is also an ornamental plant for many households around the world.

Monkshood - Aconitum (Ranunculaceae)

Aconitum has a long history of use for preparing poisons. People used it to tip their arrows and to kill wolves or poison rats. Since all the plants in this species are very poisonous, one must exercise extreme caution when dealing with them. Other names include monkshood, devil's helmet, wolfsbane, women's bane, mouse bane, blue rocket, and queen of poisons.

Description

The dark green leaves have five to seven segments in the deeply palmated lobes. Each of those segments becomes trilobed and have sharp teeth. The arrangement of the leaves is in a spiral fashion. The lower leaves have long petioles. The zygomorphic flowers are purple, pink, blue, yellow, or white, with many stamens. It has five petaloid sepals called galea shaped like the hood of a monk. The petals are 2-10 with two petals on top being large and these large petals occur under the hood of the calyx supported by long stalks.

At the apex, you see a small spur that contains the nectar. All other petals remain small or do not form. Three to five carpels get fused partially at the base.

Habitat and Distribution

These plants grow natively in the Northern Hemisphere. It prefers mountainous terrain, but you might also see them growing in well-drained mountain meadows. You can find them in the temperate zones of Canada and the United States. They also grow in parts of Europe, Africa, and Asia.

Toxicity

Everything in this plant contains toxins. Aconitine is the most potent of toxins. This potent cardiotoxin and neurotoxin cause persistent depolarization of the sodium channels. The influx of sodium through these nerve channels

Symptoms of poisoning: The heartbeat will change and become either slow or fast. Symptoms appear within a few minutes of ingesting the poison. Tingling or numbness might also manifest. Gastrointestinal manifestations include diarrhea, abdominal pain, vomiting, and nausea.

Use

People grow this perennial herb as an ornamental plant because of its bright, blue, and purple flowers. The also process the plant to use it for medicinal purposes. Therapeutic uses include treatments for joint and muscle pain. You can apply it as a tincture in cardiac patients to slow the heartbeat. Some use it for treating cold and fever symptoms.

Star of Bethlehem- Hippobroma longiflora (Campanulaceae)

This herbaceous perennial plant has the common name, Star of Bethlehem. Other names include frog's flower, horse poison, and madam fate.

Description

This perennial herbaceous species has a recumbent or erect stem. With white fleshy roots and milky sap, it grows to 20-60 cm in height. The leaves are subsessile, alternate, and simple. The shape is obovate-lancelot to elliptic with margins that are irregularly toothed. Its apex remains pointed 7-14 cm long, 1-3.4 cm wide glabrous above, a little villous beneath. The nervations are prominent. Its flowers are solitary, erect, and bloom at the axils of upper leaves on a peduncle 1.4 cm long.

We see fruits in the form of obovoid or ellipsoid bil0cular capsules. They are 1-1.5 cm long and 0.85-1.25 cm in diameter and villous. The seeds it contains are almost ovoid in shape and brown in color.

Toxicity

Ornamental but very poisonous, the sap of the plant produces high irritation immediately on contact with the skin. If it reaches the eyes, it may cause blindness. Long contact will produce fatalities. So you use gloves while handling this plant or its flower. The toxicity is due to alkaloids like lobelanidine.

Habitat and Distribution

This flowering plant is endemic to the West Indies. Of late, it grows naturally in Oceania and American tropics. It grows along sheltered banks and margins of water streams. You can also see it growing naturally in many tropical countries.

Use

In many parts of Brazil, people use it as an antiasthmatic.

Paperbark Tea Tree - Melaleuca quinquenervia (Myrtaceae)

The broad-growing tree with typical paper-like bark produces the Cajeput essential oil. It belongs to the myrtle family, Myrtaceae, and other names for this tree are the punk tree, paperbark tea tree, niaouli, and broad-leaved paperbark.

Description

The leaves are leathery and flat, 7 cm long and 2 cm wide with an alternate arrangement. They have lance to oval shapes with longitudinal and distinct veins. Their creamy white flowers appear like a short bottle brush in spikes at the end of the branches in autumn. Their petals are 3 mm long, but they fall off when the flower becomes mature. The spikes support 5-17 flower groups of threes. The tree grows 8-15 meters in height and has a spread of 5-10 meters.

Habitat and Distribution

We see the growth of the Melaleuca quinquenervia along the East Coast of Australia from Cape York to Botany Bay. It prefers swampy to silty soil, so we find it along the estuary borders. It grows alongside trees such as Bangalay and swamp mahogany. The other places it grows natively include Papua New Guinea and parts of Indonesian West Papua. It also grows in the Isles of Pines, Mare, and Belep in New Caledonia.

Toxicity

Overdoses of Melaleuca quinquenervia oil can lead to kidney inflammation, gastroenteritis, and cause disturbances in the nervous system. Sensitive persons will experience skin eruptions, even when exposed to small amounts. Their volatile nature can irritate the human respiratory system. Though Cajeput honey is good for use, sensitive individuals should exercise caution.

Use

We use the essential oil we get from its leaves for flavoring foods that include condiments, candy, baked goods, meat products, dairy desserts, relishes, and nonalcoholic beverages. If we steep the flower in water, it adds sweetness to it. We can make tea with an infusion of the leaves and flowers of this tree. To make baked food items, we use the bark as a wrapping. Also, the bark serves as a lining in hanging baskets. The bruised young aromatic leaves help treat general sickness, headaches, and cold. They also use the paper-like bark to make coolamons and shelters.

Hurricane Plant - Monstera deliciosa (Araceae)

It is a houseplant with the name the swiss cheese plant due to its quirky natural leaf holes. The vibrant green hues give the interiors an instant jungle vibe. It needs bright to medium direct-indirect light to survive. The evergreen vine goes by the names Breadfruit vine, Pinanona, Casiman, Hurricane plant, and Monstera.

Description

The broad green leaves have holes or fenestrations. One theory for the holes is that it helps improve the spread of sunlight on the forest floor while maintaining its wide disposition. It is an epiphyte with aerial roots and can reach 20 meters in height. The heart-shaped leaves are 25-88 cm long and 28-75 cm broad. Young plants have leaves that have no lobes or holes. Leaves can grow big up to 1 meter. They are heart-shaped and pinnate. When they mature, the holes appear.

The inflorescence grows to the darkest part until they find a tree trunk and grow toward the light. Its flowers are cream-white with a soft velvety appearance resembling a hood. The spadix is 10-14 cm long and about 3 cm in diameter and it is yellowish-white in color. The flowers are self-pollinating because they have both the gynoecium and the androecium.

Habitat and Distribution

This mildly invasive species grows south of Panama in the southern Mexico tropical forests. You can find it growing in the wild in Seychelles, Honduras, Costa Rica, Society Islands, Ascension Island, and Hawaii.

Toxicity

In the Hurricane plant, the toxicity happens due to insoluble calcium oxalates. It causes intense burning along with irritation in the mouth, and the person will have difficulty swallowing. The oral irritation will occur along with irritation of the lips and tongue. The person will also experience vomiting and excessive drooling.

Use

This aroid gives us edible fruits. It has leaf holes called fenestrations. But the chances of indoor plants producing flowers and fruits are very small.

Physic Nut - Jatropha curcas (Euphorbiaceae)

This semi-evergreen shrub also goes by the name of Physic Nut. This monoecious and deciduous shrub can grow to become a tree 6

31

meters tall. Other names for this plant are bubble bush, poison nut, Barbados nut, and purging nut. It belongs to the Euphorbiaceae family.

Description

The pale green to green leaves are subopposite to alternate. They have three to five lobes in a spiral arrangement. The same inflorescence has both male and female flowers. On

average, 10-20 male flowers occur in one female flower. On rare occasions, one might also see hermaphroditic flowers. Fruit production begins in late summer and goes on until fall. When their capsules change from green to yellow, the seeds are mature.

Habitat and Distribution

It is native to the American region between Argentina and Mexico. It has now spread to almost all tropical and subtropical areas in the world. It can grow in wastelands; it can grow anywhere - in sandy, saline, and rocky soil. Germination becomes complete in nine days.

Toxicity

The phorbol esters make this plant toxic. Everything in the plant is poisonous and purgative. The bark of this plant has hydrogen cyanide; people use it as a fish poison.

Use

They use the sap for marking and stain linen. At times, you can use it to blow bubbles. The latex can inhibit the watermelon mosaic virus. Sudanese use the seeds as a contraceptive.

The 4 O'Clock Flower - Mirabilis jalapa (Nyctaginaceae)

This is a common ornamental plant grown for its fragrance. The other names for this flower are 4 o'clock flower, Beauty of the Night, Garden Four o Clock, and Marvel of Peru. These flowers that blossom at 4 p.m. give a strong, sweet fragrance throughout the night and then close in the morning. It belongs to the Nyctaginaceae family.

Description

This perennial, herbaceous bush can grow one or 2 meters tall. In the temperate zone, they grow it as an annual. They are leafy and multibranched. The leaves are opposite pointed 2-4 inches long. Flowers are borne singly or in clusters. They have black carrot-shaped tubers that can weigh up to 18 kg.

Flower patterns consist of sectors, spots, and flakes, and the trumpet-shaped leaves have five petals. Colors include white, pink magenta, red, and yellow. Different combinations of flowers and colors can occur on the same plant. Another interesting phenomenon is the way it changes colors. This occurs when the plant matures.

Every inflorescence contains 7-12 un-popped flowers. It releases a scent to attract moths that help pollinate it. It remains open for 16-19 hours, allowing you to see it for a while in the morning.

Habitat and Distribution

Considered a native of tropical America, the 4 o'clock flower grows naturally in tropical, subtropical, and temperate regions of the world. In regions other than the tropics, it will die when the frost first appears. It will regrow from its bulb. It thrives in the dry tropical regions of central and South America, such as Peru, Chile, Mexico, and Guatemala. We also find it in many other countries in Europe, Middle East, Africa, America, and Asia.

Toxicity

The various parts of the plant, roots, and seeds are toxic but only if eaten. This causes stomach pain, diarrhea, and vomiting. The manifestations include skin irritation that lasts only for a few minutes.

Use

This is a much-loved garden favorite that is disease- and pest-resistant.

Yellow Oleander - Thevetia peruviana (Apocynaceae)

The common names for this include a be-still tree, Yellow Oleander, and Lucky nut. It belongs to the Apocynaceae family. This relative of Nerium oleander has a significance in religious ceremonies for Hindus in India.

Description

It is a small evergreen tree or a tropical shrub. Leaves are linear-lanceolate with a willow-like appearance. The waxy coating on the surface of the glossy green leaves helps the plant prevent water loss. Flowers bloom through summer until fall. They appear in terminal clusters in yellow and sometimes white or occasionally apricot colors. The flowers might have fragrance.

Fruits are angular drupes, green in color. Sepals fuse basally while petals have an orange, white, and yellow color. When the fruits become ripe, they turn from red to black.

Habitat and Distribution

All species of this family are native to Central America but grow in all the tropical and subtropical regions of the world.

Toxicity

Any animal or human eating this plant will get affected. There are several cardenolides, especially thevetin A and thevetin B, that cause the toxicity. Others include ruvoside, thevetoxin, neriifolin, and peruvoside and these toxins cannot be destroyed by heating or drying. They produce gastric and cardiotoxic effects.

Antidote

Treatment for poisoning includes oral administration of activated charcoal along with medicines like digoxin immune fabs and atropine. The sodium-potassium pump in the cellular membrane gets inhibited by the cardiac glycosides. This results in a drop in the electric conductivity of the heart resulting in irregular activity which can stop the heart. You must induce gastric lavage in dogs and cats if they get poisoned by this plant. Death can result in humans due to decreased cardiac output.

Use

It is an ornamental plant used in gardens and parks. In winter, it is a greenhouse or houseplant. It is drought-resistant and grows well in most soils. One can use the toxins in the plant to do biological pest control. Painting with the seed oil helps in anti-termite, antibacterial, and antifungal applications.

Rocktrumpet - Urechites Lutea (Apocynaceae)

With a neon yellow colored flower, the Urechites Lutea is an evergreen climber that we see in many gardens. Though it is not fragrant, it is much loved for its bright disposition. It has plenty of names such as wild allamanda, rocktrumpet, licebush, hammock viper's tail, yellow Mandeville, and wild Wist. This belongs to the Apocynaceae family.

Description

This evergreen climber is thorn less with a smooth stem. It has glossy leaves with poisonous sap. At the tips of the new growth and on the side branches, you see light yellow blossoms that resemble trumpets. The leaves are thick, glossy, and shiny, and the branches are thick. You get flowers in many colors, including red, yellow, pink, and white.

Habitat and Distribution

This flowering plant grows natively in southwestern United States, South America, the West Indies, Central America, and Mexico.

Too much water can kill the plant. It will survive in poor soil conditions and is quite drought-resistant. It flowers all-around the year, and though it likes full sun, direct exposure might cause a burn.

Toxicity

The action of toxicity is much like those caused by cardioactive steroids. These have a latent period that depends on the amount of substance ingested. Many kinds of dysrhythmias occur, including premature ventricular contractions, sinus bradycardia, ventricular tachydysrhythmias, and atrioventricular conduction defects, among others.

Treatment

You need gastrointestinal decontamination along with serum potassium determination and serial electrocardiograms. If serious steroid poisoning occurs, you administer digoxin-specific antibodies.

Use

People grow this garden plant on a pergola or trellis. It serves as a good decorative plant for the poolside.

Jellybean Tree - Parkinsonia aculeata (Leguminosae)

This perennial flowering tree belongs to the pea family Leguminosae. This invasive species forms dense thickets that need precise management and has several names such as Jerusalem thorn, jellybean tree, horse bean, Barbados flower-fence, sessaban, retaima, Parkinsonia, and Palo verde.

Description

This is a spiny shrub that grows up to 6 - 30 feet in height. It has graceful, drooping, hairless branches with delicate leaves. Some have a zigzag orientation. Sharp spines 3-18 mm long are present beneath each leaf. The branching is from older stems and leads to the outer whorl of spines. It bears sprays of yellow flowers with five yellow petals ⅓-⅔ inches long.

On the younger plants, the leaves are compound (pinnate), but as the plant becomes mature, these become double-compound in nature. Each of these pairs of leaves remains grouped as one to three pairs of long, flat, strap-like branchlets. These are 20-40 cm in length. The branchlets bear small leaflets without hair that are 1-9 mm long and 1-2 mm wide.

The flowers are 2-3 cm wide arranged in clusters that are 4-20 cm across and loose and elongated. These arise from leaf forks and contain 7-15 flowers, but single flowers might also be present. Flowers are bright yellow with spots of orange or red in the center. Pedicles that bear these flowers are 6-20 mm long. The reddish-yellow sepals are five in number and fuse together at the base giving five petals and 10 stamens (6-18 mm long).

Habitat and Distribution

It is native to the southern USA, Galapagos Island, northern South America, and northern Mexico. In northern South America, you find it in Uruguay, Paraguay, Peru, northern Argentina, and Bolivia inhabiting banks of water bodies, creeks, and rivers. You will also see it in grasslands, open woodlands, roadsides, and disturbed sites.

Toxicity

The leaves contain the toxin hydrocyanic acid.

Use

They plant these trees along riverbanks and water bodies to control soil erosion.

Daffodil - Narcissus (Amaryllidaceae)

One of the best flowers that can grow in any garden, Narcissus or daffodil, is one of the easiest spring-flowering bulbs to grow. The scapose plants have a single leafless stem. Daffodils look good planted on borders. Another name for the daffodil is jonquil.

Description

The plant has many basal leaves that are linear and strap-shaped or linguate. They may have a petiole stalk. Those that do are pedicellate and those that do not, sessile. The leaves have a flat and broad to cylindrical shape that arises from the bulb. When the plant emerges, it has two leaves, but when it matures, there are three or four. Their waxy appearance comes from the cutin that contains the cuticle. The color of the leaf is light-green to blue-green.

The flower is scapose. The solitary stem may have a single flower or an umbel of 20 blooms. Before it opens, the flower buds remain protected by a papery membrane called spathe. The flowers are hermaphroditic with three parts, and it is radially symmetric but might remain a little bilateral. They have a conspicuous corona trumpet. There are three separate floral patterns - Daffodil, Paperwhite, and Triandrus.

Habitat and Distribution

Narcissus is tropical or subtropical and grows natively in the Mediterranean region. This includes Spain and Portugal though some species occur in the Balkans, Italy, and southern France. One may also see them in middle and northern Europe, Great Britain, China, Japan, western, and Central Asia.

Toxicity

Lycorine, the alkaloid poison, is present in all species of narcissus. Ingesting narcissus can have a fatal effect on both animals and humans.

Use

Grind the root of the plant with honey and apply it to bruises, burns, freckles, and dislocations. You can also draw out thorns and splinters with this.

Pheasant's Eye - Adonis vernalis (Ranunculaceae)

This flowering plant grows in many locations, including dry meadows, forest clearings, and open thickets. It is a poisonous plant used as a decorative addition to household interiors and gardens. It has many names, including spring pheasant's eye, pheasant's eye, false hellebore, and yellow pheasant's eye. It belongs to the buttercup family Ranunculaceae.

Description

This herbaceous, perennial plant produces a tight cluster of stems from a short, stout rootstock. Flowers appear in springtime with as many as 20 bright yellow petals. They are 80 mm across. The stem is 20 cm tall.

Habitat and Distribution

It grows on the steppes and dry meadows of Eurasia. We also see it grow in the western regions of Spain. It grows in Switzerland, southern Europe, Sweden, and the West Siberian Plain.

Toxicity

The plant contains cardiac stimulating compounds aconitic acid and adonidin, and so it is poisonous.

Use

Bekhterev's mixture contains the infusion of the plant. They help treat edema and swelling in the body. Extracts with sodium bromide help treat panic disorders, heart diseases, and mild forms of epilepsy. Further, people use it as an ornamental plant.

Bead Tree - Melia azedarach (Meliaceae)

This tree or medium-sized shrub invades disturbed areas and is very resistant to attack by insects. Propagation is with the help of birds that eat the fruits of the tree and drop the seeds at distant places. It has plenty of other names such as syringa berry tree, Cape lilac, bead tree, Pride of India, chinaberry tree, white cedar, Indian lilac, and Persian lilac. It belongs to the Meliaceae family.

Description

When the tree is grown, it has a rounded crown. It grows up to 7-11 meters on average but can also reach a height of 45 meters. The leaves of this tree are 50 cm long and two or three times compound. They are long-petioled and alternate. The color of the leaf is lighter green below and dark green above. The margins have serrations and the flowers have five pale lilac or purple petals. They are small and fragrant and grow in small clusters.

The fruit is marble-sized and light yellow in color. All winter, it stays on the tree. It becomes wrinkled in time and loses color to become white.

Habitat and Distribution

It grows in Sri Lanka, India, southern and central China, Laos, and Thailand. You can also see it grow in eastern Australia, Philippines, Indonesia, Vietnam, Bhutan, and Nepal. It grows in moist sunny locations up to 2,700 meters in the Himalayas.

Toxicity

Most parts of the tree, bark, flowers, and leaves are poisonous. At times, even the fruit is poisonous. The flowers produce respiratory irritation. The person might die within 24 hours of eating the fruit of this tree. The toxicity is within the seed. So, if the person crushes the covering and the seeds while eating, he will die. Birds do not feel the effects of the poison and so eat plenty of them and "get drunk".

The symptoms of poisoning occur within hours of eating the fruit. These include vomiting, loss of appetite, general weakness, lack of coordination, and rigidity. The tetranortriterpenoids, related to the insecticidal compound azadirachtin, cause toxicity. The leaves form a natural insecticide, but one must not eat them.

Use

A diluted infusion made from the parts of the tree and leaves helps in inducing uterus relaxation. Leaves are a natural insecticide within which people keep food. They place Chinaberry fruit inside drying apple crates to prevent the formation of insect larvae.

Carolina Bloodroot Lachnanthes tinctoria (Haemodoraceae)

It also has the name Spirit Weed. Other names of this plant are Carolina bloodroot and a red root. Medically, people refer to the preparation from this plant as mother tincture. It belongs to the Haemodoraceae family.

Description

The plant grows 32-52 inches and prefers the sandy soil of the coast and swamps such as Florida and Carolina in the United States. The flower has six tepals of pale yellow color. It has red roots and rhizomes. With an erect, vertically ribbed, and jointed stem, the plant has whorled leaves that are minute and remain whorled into a sheath. It has terminal teeth, and the cones are terminal.

Habitat and Distribution

Lachnanthes tinctoria is native to southeastern Nova Scotia, eastern North America, south of Cuba and Florida, Massachusetts, and along the Gulf of Mexico up to Louisiana. You can also see it in some parts of the western Caribbean islands off Honduras.

Toxicity

Horses become poisoned by this plant. Symptoms are the same as those seen in bracken fern except that there is no change in the appetite until the terminal stages.

There are an increase and weakening of the pulse rate with nervousness as an early sign. Difficulty in turning is present. In animals, there is rigidity, and constipation might occur. The cornea of the eye might become opaque. Calm coma precedes death.

Use

Medicinal preparation of the parts of the plant help cure tuberculosis, stiff neck, sore throat, typhoid, pneumonia, and diphtheria.

Gloriosa superba (Gloriosa)

Usually, climbing, this perennial shrub, is very toxic. The plant grows well in the tropics and subtropical regions in the lowlands. Sunbirds and butterflies pollinate it. It can grow in nutrient-poor conditions, so we find it growing not only in woodlands but also in sand dunes and grasslands.

Description

This perennial shrub has a fleshy rhizome. Due to its scandent nature, the stem can grow up to 4 meters using modified leaf tip tendrils. Leaves are alternate, but on occasions, it might also be the opposite. Tips have tendrils, and shape is lance-like with a length up to 13 - 19 cm. Flowers have six showy tepals 5.6 - 7.6 cm long in bright red and orange colors at maturity. The base is a little yellow, and margins are wavy. The six stamens are 4 cm long, and it has large amounts of yellow stamen at its tip. The fruit is fleshy and is 6-12 cm long.

Habitat and Distribution

It grows in tropical regions and is native to southern Africa. You can also see them in tropical Asia and regions along the Indian Ocean. They grow in Indonesia and Malaysia. They prefer to grow in hedges and open forest at the height of 300 meters.

Toxicity

The alkaloid colchicine gives this plant its toxicity. The signs of poisoning are abdominal pains and vomiting. In the latter stages, diarrhea will become severe, and bleed will occur. The person will experience high blood pressure and go into shock. Dehydration and metabolic acidosis will occur. Severe reactions will show vascular damage, hematuria, and oliguria. Death happens due to cardiovascular collapse and respiratory depression.

Use

Used in traditional medicine in Asia and Africa. People also use it in plant breeding to produce polyploidy. They also use the extracts of shoots and tubers that show nematicidal activity.

Golden Dewdrop - Duranta repens (Verbenaceae)

Geisha Girl or the Duranta repens is a versatile, colorful plant with a vanilla scent. The berries and leaves of this plant are toxic. It has a very fast growth rate but grows up to medium height only. Other common names for this plant include golden dewdrop, skyflower, Sapphire shower, Alba, Aussie Gold, and pigeon berry. This plant belongs to the verbena family.

Description

This sprawling shrub can grow into dense thickets if not pruned in time. They grow up to a height of 6 meters. Mature plants will show axillary thorns that are not present on the younger plants. They also bear glossy, golden fruits 5-14 cm wide. Leaves are opposite, ovate to elliptic with light green color.

They are 7.3 cm long and 3.4 cm broad, and their petioles are 1.5 cm long. Flowers are lavender to light-blue. You can see some spines on the stems with one on each base of the leaf stalk. The leaf blades have a length of 17-85 mm and 13-57 mm wide with entire margins and they might have serrations toward the round or pointed tips.

The flowers have short stalks and bloom as elongated clusters 4-27 cm in size. Each flower has a thin tube 1 cm and has fused petals with five distinct lobes. There are five sepals 3-9 mm long, and the flowers are 8-17 mm long. The fruits are round and borne as large clusters.

Habitat and Distribution

It grows natively in the southern USA regions of Texas and southern Florida, the Caribbean, and regions of Central America such as Panama, Nicaragua, Honduras, Guatemala, El Salvador, Costa Rica, and Belize. In South America, you can see it growing in Brazil, Surinam, Venezuela, Paraguay, Argentina, Peru, Ecuador, Colombia, and Bolivia.

Toxicity

The leaves and berries are poisonous to both animals and humans. Poisoning in dogs and cats manifests as drowsiness, tetanic seizures, and hyperaesthesia along with alimentary tract irritation. This will show up as diarrhea, bleed, and vomiting.

Use

It serves as an ornamental plant in most regions of the world. Golden Dewdrop possesses antimicrobial and insecticide properties. The therapeutic properties are due to the presence of glycosides, flavonoids, alkaloids, saponins, tannins, and steroids besides other useful compounds.

Wedge-Leaf Rattle Pod - Crotalaria retusa(Fabaceae)

Poisonous to livestock, the Crotalaria retusa also contaminate human food. This annual plant grows along rivers up to 250 meters elevation. In deciduous forests, you can find them at heights of 1.500 meters. Common names include wedge-leaf rattle pod, shak shak, Yellow Lupine, rattle weed, and devil-bean. They belong to the Fabaceae family.

Description

This annual herb has an erect stem and grows 1.25 meters tall. It has simple inverted lance-shaped leaves 3.3-9.4 cm in length and 1.4-3.9 cm wide. Tips may have points but are usually round and notched. The base of the leaf is wedge-shaped, and stalks are 3.2 mm long. Flowers are 1.4-2.7 cm long and 12-23 cm wide. They have lax spines at the end of branches that are up to 27 cm when you include the stalk. The yellow petals have a purple line near the base. These are 1.4-1.8 cm long and 1.6-2.4 cm wide with oblong-lance shaped wings. Pods are greenish and then turn to brown and eventually become black. They are hairless and 3.4 cm long.

Habitat and Distribution

It is native to Kimberly in Australia. The Crotalaria retusa prefers clayey soil. We also find it growing in Texas, northern Mexico, eastern Arizona, and southern New Mexico.

Toxicity

Swainsonine is the toxic principle here. It is an alkaloid and causes the poisoning of cattle, goats, sheep, and horses. They need to eat 90% of their bodyweight before showing any signs of poisoning for horses. This is because there are pyrrolizidine hepatotoxic alkaloids. Eating 200% of the body weight for 2-3 months will kill sheep, goats, and cattle. The horses and cattle begin to walk aimlessly until they die.

Use

The sweet flavor of the flowers and leaves make it a preferred vegetable in many parts of the world. Medicinally, they use it to soothe cold by making a decoction of the leaves and flowers. It is also used to treat many other complaints such as fevers, cardiac disorders, impetigo, scabies, and diarrhea. They use an infusion of the plant for bathing children to fight against skin infections. Raw seeds act as an analgesic for lessening the pain of a scorpion sting. At times, they use it as a fiber crop.

Spider Lily - Crinum asiaticum (Amaryllidaceae)

The large Crinum asiaticum plants love heat and moisture and have poison characteristics of low severity. These summer-flowering bulbs have statuesque and lustrous foliage. The other names for this plant are spider lily, grand crinum lily, poison bulb, giant crinum lily, and swamp lily. They belong to the Amaryllidaceae family.

Description

This perennial herb with strap-shaped leaves will grow to over 5 feet in a short time. Its spherical pseudobulb that forms the leaf base is cylindrical, and base branched laterally. Its diameter is 6-15 cm.

Leaves are 5 feet long, lancelot with margins that undulate. The dark green single apex is acuminate up to 1 m long and 7-11 cm wide with 20-30 leaves in each cluster.

The aromatic inflorescence is umbel with 11-23 flowers with multiple petals. The solid stem of the flower is erect and as long as the leaf. It has 6 reddish stamens and corolla lobes. The fruit is green and 3-5 cm across as an oblate capsule.

Habitat and Distribution

The Crinum Asiatica is native to East Asia, Indian Ocean islands, Pacific Islands, Australia, and tropical Asia. You can see the naturalized flowering plants in Florida, the

West Indies, Mexico, Chagos Archipelago, Madagascar, many Pacific islands, and Louisiana.

Toxicity

All things in the plant are poisonous if you eat them. But the toxicity level is low. Exposure to the sap will cause skin irritation.

Use

People use the leaf in an external application as a decoction. It helps deal with swelling toxicity, laryngopharyngitis. Adenolymphitis, headache, numbness, arthralgia spasm, fractures, and bruises. One may use the extract from the bulb, fruit, or leaf.

Queen of Climbers - Clematis (Ranunculaceae)

These vigorous, woody, climbing vines have a fragile stem when they are young. People grow the "Queen of Climbers" on trellises and walls during summertime to add variety and color to the garden and open spaces. There are also evergreen winter-flowering varieties and common names include traveler's joy, old man's beard, virgin's bower, vase vine, and leather flower. They belong to the Ranunculaceae family.

Description

The clematis takes up to three years to develop a strong root system. So, growth will remain slow during the first two years. Leaves are opposite and comprise leafstalks and leaflets that curl around structures for support. Those varieties that grow in cool climates are deciduous, while those in warmer climes are evergreen.

Flowers come in a variety of shapes, depending on the species. They include star-shape, semi-double, double, single, tubular, tulip-shaped, bell-shaped, and open bell-shape. Most of them are less than 3 inches in size, while a few can be as big as 8 inches across. The color ranges from red, white, and deep purple to pink and blue.

Habitat and Distribution

We can find this growing natively in temperate regions of the Northern Hemisphere. They rarely grow in tropical regions.

Toxicity

Plants of this genus contain compounds and essential oils that are very irritating to the mucous membrane and the skin. This substance is very toxic and can cause bleeding in the digestive tract if ingested in large amounts.

Use

Used in small amounts, the extract of the plant can cure migraine headaches. It will also help treat nervous disorders. People also use it to treat skin infections. The extracts show antifungal and antibacterial activity. You can use them for anti-inflammatory and wound healing activities.

Golden Shower - Cassia fistula (Fabaceae)

This ornamental plant is the National Flower of Thailand. You can find it in many gardens and public places because they make the surroundings beautiful. The bright yellow color of the flowers adds to the splendor of any environment. Other names for this plant are the pudding-pipe tree, golden shower, Indian laburnum, and purging cassia. This beautiful tropical tree also has the name of Amaltas. It belongs to the Fabaceae family.

Description

This ornamental tree with a hard, red trunk can grow up to 40 feet in height. The smooth, slender, pale-gray stem bark turns dark brown and rough when the tree ages. The spirally arranged leaves are alternate and paripinnately compound. The 30-40 cm long leaves have a pinnate arrangement of 3-8 ovate paired leaflets 7.5-12 cm long and 2-5 cm broad. They occur on terminal racemes that droop that are 30-60 cm long. The little zygomorphic, showy flowers are bright yellow and pentamerous.

The fruit has the shape of a cylindrical, pendulous pod 60-95 cm long and 1.5-2 cm wide. The ellipsoid seeds are 8.9 mm long and glossy brown.

Habitat and Distribution

Though native to Southeast Asia, this tree occurs in all tropical regions. We find it growing natively on the Indian subcontinent. It spreads eastward to Myanmar, and Thailand, and southward to Sri Lanka. Cultivation of the tree occurs in French Guiana, Guyana, Belize, Ecuador, Mexico, the West Indies, and Costa Rica.

Toxicity

There were no significant toxic effects on the rats used in the study. Traditionally, people used the extract as a laxative because of the anthraquinone glycoside content. In this, the major anthraquinone is Rhein. There was no mortality in rats when the toxicity levels increased.

Use

Cassia fistula is an ornamental plant for many purposes. The tree gives good quality charcoal and fuel along with heavy and hard timber. They use it to make wheels, posts, farm implements, and furniture. We also get dyestuff and tannins from the bark. Twigs of the Golden Shower tree are excellent fodder material. Flowers and leaves give extracts with antibacterial activity. The pulp of the seedpod is a mild laxative.

Elephant Ears - Caladium (Araceae)

The Heart of Jesus plant, Caladium, has showy and colorful, heart-shaped leaves with color combinations ranging from red and white to pink and green. They thrive in hot, humid weather conditions and prefer full to partial shade conditions. They have prominently colored midribs, contrasting margins, and patterns, including striped and mottled veins. The common name for this plant is Elephant Ears. People also call it the Mother-in-law and Angel's Wings plant. It belongs to the Caladieae family.

Description

The leaves occur on long petioles as there are no stems. There are two main types of shapes for the leaves - full heart shape and semi-heart shape. The first, fancy-leaf variety is more common. Plants grow wild up to a height of 40-90 cm with leaves 16 cm long and 47 cm wide. Fruits are berries with many seeds.

Habitat and Distribution

The Caladium is native to South and Central America. They grow on the banks of rivers and in open forests and they become dormant during the dry season. In India, many coastal islands, and parts of Africa, it grows as an introduced species.

Toxicity

Everything in the Caladium is toxic when you chew or swallow this plant. Symptoms include painful and intense burning with swelling of the throat, tongue, mouth, and lips. You will also see intense salivation. Gastric irritation is present along with intense itching, which can lead to dermatitis.

Use

People use the Caladium as the borders for a walkway or beautify a corner of the garden. This reliable summer bedding plant is the ideal answer to beautify the spaces around your house. They powder the tuber to treat facial skin blemishes. Bathe children in freshwater to which you have added macerated Caladium leaf to cure maladies. The enema made from the juice of stems helps expel roundworms. You can expel the vermin on the soles of cattle using crushed leaves.

Belladonna - Atropa belladonna (Solanaceae)

This tall, bushy herb is very poisonous whose roots and leaves give us medicine. Called Belladonna or "Beautiful Lady," ladies in Italy used this to enlarge their pupils. But, this practice was dangerous. The nightshade family Solanaceae includes eggplant, potato, and tomato. Another name for this plant is deadly nightshade.

Description

This is a branching perennial herbaceous shrub growing from a fleshy rootstock. Plants achieve a height of 2 meters and ovate leaves are 18 cm long. The dull purple flowers have a faint fragrance and are bell-shaped and have green tinges. Fruits of this plant are 1.4 cm berries, green on young plants, and brown when the plant matures. Berries contain a toxic alkaloid, but the animals consume them and disperse their seeds all over the place.

Habitat and Distribution

Atropa belladonna grows natively in North Africa, southern, eastern and central Europe, the Caucasus, Iran, and Turkey. You will also see it growing in Britain, Sweden, and parts of North America.

Toxicity

Since Belladonna is one of the most toxic plants around, consuming any part of the plant will give rise to many problems such as cardiovascular diseases, complications during pregnancy, psychiatric, and gastrointestinal disorders. The root contains a huge amount of poison.

The toxin contains active ingredients scopolamine, atropine, and hyoscyamine. The symptoms of poisoning include sensitivity to light, dilated pupils, headache, staggering, loss of balance, blurred vision, and slurred speech. The toxin disrupts the nervous system and prevents it from controlling involuntary activities.

Use

Women use the extract of the belladonna plant to make their pupils huge, as this makes them look attractive. People use this as a dietary supplement. Another use is to treat colds.

Mexican Poppy - Argemone mexicana (Papaveraceae)

This poisonous hardy plant is tolerant of poor soil conditions and drought. The bright yellow latex it produces is poisonous to grazing animals. The annual plant has prickly stems and leaves. Other names are cardo/cardo Santo, flowering thistle, Mexican prickly poppy, and Mexican poppy and it belongs to the Papaveraceae family.

Description

The prickly poppy is a branched, erect, annual/perennial plant. Its first leaves are simple, green-ribbed, and arranged in a rosette and they are also white, sessile, and alternate. The blade is 6-8 cm long and 1 cm wide and ends in a short spine. The pale, greenish stem is smooth and oblong to cylindrical and the leaves are simple, sessile, and alternate and blue-green color. They are blue-green in color, thick, and leathery. Their shape is obovate, pinnate, lobed 6-20 cm long and 3-8 cm wide.

It has 4-7 cm sized flowers, solitary sessile, terminal or axillary, and bright yellow and has 6 round, bright-yellow petals. Fruits are ovoid capsules, divided into five chambers. There are many seeds, each 1.72-2 mm by 1.5 mm.

Habitat and Distribution

It prefers dry soils, and you can see them growing on the roadsides. It grows natively in tropical America. In Africa, it grows in the north, east, and far south.

Toxicity

Argemone mexicana resembles mustard seeds, and so there is the danger of adulteration. Everything about the plant is toxic, including the seeds. The cattle get poisoned by eating the dried leaves, and poultry has many fatalities eating the seeds. The oil of the seed is toxic, causing intense pain all over the body. It also causes diarrhea and fever.

Use

This plant is a wonderful decorative plant. People use this plant extract to treat warts and wounds.

Flamingo Lily - Anthurium (Araceae)

The outstanding feature of the Anthurium is their brightly colored flower spathe. It is a popular foliage plant well-suited to brighten the interiors of a home or retail outlet. Anthurium flowers are very clean and do not dirty up the water you put them in. All parts of the plant are poisonous if ingested. The other names of this plant include Tail Flower, Flamingo Lily, Laceleaf, and Painted Tongue and it belongs to the Araceae family.

Description

Anthurium leaf shape is variable and is often clustered. They might be epiphytic or terrestrial. Small male and female flower structures occur in the inflorescence. The flowers are on the spadix in dense spirals. The spadix itself remains elongated like a spike, but it might also be club or globe-shaped. The spathe lies beneath the spadix and is lance-shaped in many species. It might have a curve or remain flat while covering the spadix like a hood. Fruits are juicy berries containing two seeds.

Habitat and Distribution

This plant is native to the Americas. You find it growing in the Caribbean and from northern Argentina to northern Mexico.

Toxicity

These plants are poisonous due to the presence of calcium oxalate. If you touch the sap, it will irritate the eyes or skin.

Use

Used as an ornamental plant, they propagate the Anthurium by seeds or through cuttings.

Amaryllis (Amaryllidaceae)

Amaryllis is a wonderful indoor bulb easy to grow and available in many colors. They do not need any special treatment and grow indoors well. Common names for the flowers in this genus include Jersey lily, belladonna lily, Easter lily, Amarillo, and naked lady. They have the name lily because of their shape and habit of growth. This plant belongs to the Amaryllidaceae family.

Description

The size of the bulbs is 5-10 cm, and the size of the strap-shaped green leaves is 32-52 cm length and 2-3 cm width in two rows. Two leafless stems emerge from each bulb and grow up to 30-60 cm. At the top of these stems, we see 2-12 funnel-shaped leaves. The flower is 6-10 cm in size with six tepals arranged in two rows, three in the inner and three in the outer row. The usual color is white with purple veins. Other colors are pink and purple.

Habitat and Distribution

This plant is native to the warm regions of South America and is also seen in the Caribbean. The genus is native to South Africa also.

Toxicity

All species in this genus are toxic to humans if ingested. Symptoms of poisoning include excess saliva, abdominal pain, tremors, vomiting, diarrhea, abdominal pain and long-term effects of poisoning include loss of appetite and depression.

Use

People use it as potted plants to decorate the interiors. Bright, indirect sunlight helps indoor plants, while partial or full shade helps outdoor plants.

Night Blooming Jessamine - Cestrum nocturnum

The white tubular flowers of this plant bloom at night, releasing the scent of perfume that spreads to 300 feet distance or more. This makes it the plant with the strongest scent. The other names for this plant are Night Jessamine, Lady of the Night, Night Queen, Galan de Noche, Night Blooming Jessamine, Night Jasmine, Queen of the Night, Night Blooming Cestrum, and Dama de Noche. It belongs to the Solanaceae family. That is also known as the potato or nightshade family.

Description

The stems are pubescent (have fine hair) and flexuous (they bend and twine) with smaller twigs displaying more hair. The elliptical, long, and lanceolate leaves are 10-14 cm long and 4-7 cm wide. They are glossy and smooth and have even margins. You find fine hair on the undersurface of the midrib.

The flowers occur at the end of the branches in dense clusters. They emerge at the junctions of leaves and twigs which results in a dense crowd of leaves and flowers. The tubular, greenish-white flowers split into five sharply pointed petals or triangles at the top of the tube.

The tubular part of the flower is 2-2.5 cm long. Once opened at night, the flower is 1-1.3 cm across. The floral tube contains the anthers and stamen. Fruits are berries, green at first, then turns white.

Habitat and Distribution

We see this growing on many islands of the Pacific. It is naturalized in many tropical and subtropical regions of the world including New Zealand, Australia, the southern United States, southern China, and South Africa.

Toxicity

All plants in the Solanaceae family are toxic due to the presence of solanine. This is an alkaloid that causes respiratory problems if you smell the scent. If you ingest it, you will experience feverish symptoms. You might experience a headache, throat and nose irritation, and nausea. It is due to the chlorogenic acid present in some varieties.

Use

Night Jasmine is a houseplant grown for its sweet fragrance. It makes a good hedge plant in between houses.

St. Vincent Plum - Gliricidia sepium (Fabaceae)

This small shrub or tree has an open crown. People grow it as a support and shade tree. It has many names including quick stick, St. Vincent plum, Gliricidia, mother of cocoa, Mexican lilac, Nicaraguan cocoa shade, tree of iron, and Aaron's rod. It belongs to the Fabaceae family.

Description

The small tree that grows up to 10-12 m has a smooth bark. The whitish-gray color can change to a deep maroon color. Flowers bloom at the end of the branches that do not have any leaves. The

71

tree pod 10-15 cm long is green when young and becomes yellow-brown when mature. It has 4-9 brown seeds.

It flowers on and off many times in the year. They have a lilac or bright pink flowers arranged in racemes that are clustered. The fruit pods are 10-18 cm long and 2 cm wide with 8-10 seeds.

Habitat and Distribution

It grows natively north to Mexico and Costa Rica in Central America. It also grows in northern South America.

Toxicity

The powdered bark, seeds, and leaves are toxic to humans. Leaves are toxic to horses and dogs. Goats and cows are not affected. There is limited evidence of toxicity for normal feeding conditions for cattle.

Use

People use it as a shade tree for plantation crops such as coffee. Today, it serves many other purposes such as firewood, fodder, live fencing, rat poison, intercropping, and green manure. During paddy cropping, gliricidia leaves are plowed into the soil. It also serves to stabilize the soil. To poison rats, the powdered bark is mixed with rice and left for rats to eat.

Wild Poinsettia

Known for its red and green foliage, the poinsettia is an erect annual with a strong unbranched stem. These are shrubs or small trees that are very toxic, and so pets and children must not get close to them. They also go by the name of Fire on the Mountains, Kaliko plant, milkweed, painted spurge, painted leaf, and painted euphorbia. It belongs to the spurge (Euphorbiaceae) family.

Description

The alternate, long-stalked leaves are variable in shape. It might be oblong with lobes or very linear and some teeth. Near the inflorescence, some of the leaves will be opposite on very short stalks. It has a medium green color on the upper side, and on the lower side, it is a paler shade with fine whitish hair. At the top of the stem, you find the inflorescence. The Euphorbia flowers are unique in that the flowers are grouped inside a cyathium. Nectar glands form at the base with outward flaring appendages that make the flower look like a cup. It is only ⅛ inch in size. We find the stalk of the cyathium arising from the center of the cyathium. It has an ovary with three lobes, and three divided styles emerge. It has three oval brown seeds.

Habitat and Distribution

It occurs in rocky glades and sand prairies, along streams in the gravel bars, rocky woodlands, and roadsides. Wild poinsettia is scattered all over Illinois. It prefers disturbed areas. You can also see it in southern Florida, Mexico, and throughout the southern United States. You find it grown as an ornamental plant in Southeast and South Asia.in Thailand and India, it is considered a weed.

Toxicity

People who are sensitive to latex will show strong reactions such as anaphylaxis and dermatitis. The latex juice might also have carcinogenic effects. The toxicity is thought to be due to a resin.

Use

People use it in floral displays during Christmas due to its bright red color. Though people know it is toxic, they use it for various medical applications, including breast milk production, treating a fever, and cause abortion. It also helps induce vomiting, kills bacteria, and pain.

Devil's Weed - Datura Stramonium (Solanaceae)

Well known in medicinal herbs circles, Datura Stramonium is poisonous if you ingest a large amount. This widespread weed is present in more than 100 countries. The common names are devil's weed, devil's trumpet, jimson weed, common thorn apple, moonflower, hell's bells, devil's cucumber, prickly burr, datura, locoweed, stinkweed, and Jamestown weed. It belongs to the Solanaceae family.

Description

This annual herb is freely branching with a foul smell. It grows up to 64-145 cm in height with a stout, erect stem. The root is white, fibrous, and thick and the leaves are smooth, yellow-green, toothed, 8-20 cm long, and irregularly undulated. The flower is solitary and trumpet-shaped with five white or lavender lobes on top. They are fragrant and open at night.

The seed capsule is egg-shaped bald or covered with spines having four chambers filled with seeds.

Habitat and Distribution

Jimson weed favors drier climate and sunlight. So, you can find it in barnyards, pastures, fields, waste grounds, railroads, roadsides, and cultivated areas. It is native to North America. You can also see it in central and South America.

Toxicity

All Datura plants have very high levels of tropane alkaloids including hyoscyamine, scopolamine, and atropine. They are anticholinergics or deliriants. The risk is due to overdosing. Many who use it for recreational purposes get poisoned. Datura intoxication will produce tachycardia, hyperthermia, hallucination, delirium, severe mydriasis, urinary retention, and bizarre behavior.

Antidote

For severe cases of poisoning, intravenous physostigmine will prove useful.

Use

Datura serves as medicine, and people smoke recreationally also. Historically, they used it as an analgesic while setting bones. For the Chinese, it was an analgesic to use during surgery.

Allium (Amaryllidaceae)

There are hundreds of species in this group, and they include garlic, onion, chives, leek, shallot, and scallion. They described Allium first in 1753, and before that, the Greeks have made references to the smell of garlic. Allium is the sole genus of the Allieae that occurs in the Amaryllidaceae family. There is considerable polymorphism because they adapt to different growing conditions. Alliums grow in many shades of pink, blue, red, and white, which is why so many people grow them in their gardens. They belong to the Amaryllidaceae family.

Description

All plants in this family have underground bulbs. The strap or lance-shaped leaves grouped at the base of the stem alternate along the stem. Flowers have three to six petals and as many sepals and the fruits are fleshy berries or dry capsules. Many ornamental flower gardens feature Amaryllidoideae, the largest subfamily of this group. Daffodil, snowdrop, and belladonna lily are some examples of plants in this group.

In the subfamily Allioideae, we see many food crops such as garlic, onion, chive, and leek. African Lily is one of the plants in the third group Agapanthoideae in this family.

Habitat and Distribution

The 73 genera of plants spread across 1,600 species, prefer the tropical and subtropical regions. These plants are native to North America and Central Asia. You also find different varieties of this family growing in Australia, South America, and South Africa.

Toxicity

Humans can consume Alliums without worry, but they are toxic to cats and dogs.

Many of these species of plants find use as an ornamental border plant in the garden.

Croton betulinus (Euphorbiaceae)

This member of the spurge family produces extensive flowers. Common names for this plant include croton and rushfoil and the range of plants extends to shrubs and trees. It is also known as beech leaf croton. It belongs to the Euphorbiaceae family.

Description

This shrub or small tree includes many varieties with different characteristics. Due to their mutative abilities, they often become invasive. Young growth is sub-glabrous to sparingly pubescent. The leaf length varies with leaf blades 10-45 cm long and 1.5-10 cm wide. They are variable in shape obtuse to acute. At the apex, it is sometimes aristate. It becomes rounded cuneate to attenuate at the base. Each cultivar has a different color and flower pattern.

Habitat and Distribution

You see this species growing in the Caribbean and Florida. It grows natively in North America, coastal parts of northern South America, southeastern Africa, southeastern coast of India, and eastern coast of Australia.

Toxicity

Accidental poisoning by ingesting the croton leaves are rare because of the bitter taste it has. If ingested, you will see diarrhea and vomiting. If one eats too much, the result could be fatal.

Use

The oil of this plant is a violent purgative. But due to its toxic nature, people no longer use it. It serves as an ornamental garden plant in many regions of the world. It also helps to heal lesions, constipation, and as a liquid bandage. The bark of one species helps give flavor to liquor. It is also used as a windbreak to fight desertification.

Caryota mitis (Arecaceae)

In this genus of palm trees, clustered stems and purple flowers abound. One species gives us the much-loved jaggery. They also make palm wine from the same species. It prefers disturbed wooded regions. It belongs to the Arecaceae family.

Description

It has the common name clustering fishtail palm. With clustered stems up to 32 feet and 15 cm diameter, Caryota mitis has 10 feet long leaves and purple flowers. The fruits are toxic and colored deep red or purple. When they grow older, the palms get ringed with widely spaced leaf scars. Leaves are bipinnate, induplicate with dull, green pendulous leaflets, while the leaflets are praemorse and obdeltoid. The pendant branches have both pistillate and staminate flowers.

Habitat and Distribution

This plant prefers warm temperate to tropical zones. It is native to Indochina and prefers the mountainous regions. You can see it in Vietnam, Laos, Cambodia, Thailand, Myanmar, India, Philippines, Indonesia, and Malaysia.

Toxicity

There is a toxin in the fruit called crystal oxalate. It irritates the eyes and the skin.

Use

It is a border plant for many gardens and walkways. This ornamental plant adds to the beauty of any garden. People also burn the hair on the leaves of the plant to cure disorders of the limb.

Giant Taro - Alocasia macrorrhizos (Araceae)

The galaxy marbled tones of the Alocasia mycorrhizos will brighten up any garden space. It has the names giant taro, ape, Piya, biga, and giant alocasia. This decorative plant is toxic. In Australia, it goes by the name of cunjevoi. Other names for this plant include Giant Taro, Upright Elephant Ear, and Giant Elephant Ear. It belongs to the Araceae family.

Description

This succulent, herbaceous plant can grow up to 4.5 meters. Its arrow-shaped leaves have rounded and shallow lobes. The leaves point upward, forming a line with the main axis of the petiole. Leaves are almost peltate and green and have a conspicuous midrib. The upright stem holds the leaves upright and the flowers are insignificant but fragrant, remaining enclosed inside a yellow-green swathe.

Habitat and Distribution

It grows natively in Queensland, New Guinea, and the Island Southeast Asia rainforests. People also grow it in many Pacific islands, the Philippines, and other places in the tropics. Alocasias need warmth all the time. It also needs rich but well-drained soil. Water it daily, but make sure the water doesn't stagnate.

Toxicity

It has raphides of calcium oxalate crystals that irritate the skin. If ingested, it can cause severe stomach pain and inflammation of the tongue and mouth. You could experience sore throat chest tightness, swollen lips, dysphagia, and airway obstruction.

Use

If you cook it for a long time, you can eat it. The rootstock is a mild diuretic and laxative. It can treat diseases and inflammations of the skin and abdomen. In India, it helped in the treatment of scorpion sting. The stem of the plant helps ease the pain due to rheumatism and gout.

Cape Periwinkle - Catharanthus Roseus (Apocynaceae)

Endangered in the wild, it grows in gardens and parks in most places in the world. Its seeds are dispersed with ease by ants, water, and wind. Common names are rose periwinkle, Catharanthus roseus, cape periwinkle, rosy periwinkle, vinca, pink periwinkle, old maid, bright eyes, and Madagascar periwinkle. It belongs to the Apocynaceae family.

Description

The trees, vines, and shrubs in this group have milky latex. It is an evergreen subshrub that grows up to 1 meter in height. Leaves are oblong to oval 2.6-9.2 cm long and 1.2-3.4 cm broad. Leaves are glossy, hairless, and green arranged in opposite pairs. They have a short petiole 0.9-1.8 cm long red or green with a pale midrib. The center of the flower is dark red; petals are white to dark pink. The basal tube is 2.5-3.1 cm long and has a corolla 2-5 cm diameter. It has five petal-like lobes. The fruit is two follicles 2-4 cm long.

Habitat and Distribution

This is native to Madagascar but is now an endangered species. It grows in open forests, scrubland, sandy locations along the coast, dry waste places, savanna vegetation, and riverbanks. You can also see it grow at times on rocky soils.

Toxicity

People cultivate it for its medicinal value. They use extracts of the shoots and roots (it is poisonous) in traditional Indian medicine, Ayurveda, to treat many diseases. These include malaria, diabetes, Hodgkin's lymphoma, and leukemia.

Use

It is a medicinal and ornamental plant and is a source of the drugs vinblastine and vincristine used for treating cancer. People grow it along walkways as a decoration.

Bahama Whitewood - Wild Cinnamon (Lauraceae)

We get this spice from the inner bark of the genus Cinnamomum. This evergreen shrub is salt tolerant, and flowers in summer and produces fruits in fall. The timber of this tree goes by the name of Bahama whitewood. One may get this at any local nursery. This belongs to the Lauraceae family.

Description

It has white and purple showy flowers that cover the tree in summer. Cinnamon grows on evergreen trees that have oval-shaped leaves, berry fruit, and thick bark. Only small pieces of 0.5 mm thickness get harvested, and the outer woody bark gets discarded. This leaves the meter-long cinnamon to curl into smaller pieces.

This small bushy tree produces flowers for most of the year. The flower has white petals and is 2.5 cm in size. It has alternately arranged oval-shaped leaves. Pollen sacs are terminal on the stamens and flowers have three sepals and 4-12 petals. There is only one aperture for the pollen. About 2-6 carpels fuse to become a single locule. Ovules remain attached to the inner surface of the locule wall.

Habitat and Distribution

Cinnamon is naturalized in all countries in the world. It grows natively in tropical America and Florida in damp, shady regions.

Toxicity

Coumarin in the cinnamon will damage the kidney and liver if ingested in excess. One should have only 0.1 mg per kg of body weight to be perfectly safe always.

Use

It has a use as a condiment and you can make many sweet and savory dishes, snack foods, traditional foods, tea, and breakfast cereals. People also use it for its medicinal value. It helps flavor alcohol. You can use it to poison fish, toxic to chicken, and repels cockroaches.

Indian Laurel - Calophyllum (Calophyllaceae)

This is a very important multipurpose tree. Calophyllum grows in warm temperature and moderately wet conditions. The tamanu oil it gives is very precious. The other names for this plant include red poon, laurelwood, Indian laurel, Indian doomba oil tree, Borneo mahogany, beach Touriga, Alexandrian laurel ball tree, tacamahac tree, and satin Touriga. It belongs to the Calophyllaceae family.

Description

This is a slow-growing tree with low branches. It has irregular and broad crowns and reaches heights of 8-20 meters. It has 25 mm wide flowers in paniculate inflorescences or as racemose with 4-15 flowers. Flowering is perennial. The fruit is a green drupe 2-4 cm in size with one large seed. When ripe, the skin gets wrinkled, and the color changes to brownish red or yellow.

Habitat and Distribution

The shrubs produce yellow or white latex. It grows natively in Africa in Mauritius, Madagascar, Tanzania, and Mozambique. In Asia, we see it in China, Cambodia, Andaman and Nicobar Islands, India, Indonesia, Thailand, Philippines, Myanmar, Malaysia, and Japan. In the Pacific region, it grows in Micronesia, Cook Islands, Guam, French, and Palau. Polynesia, and in Australia in Queensland and Northern Territory.

Toxicity

The sap of the tree is poisonous. People use it to poison the fish. In Samoa, they use this poison to tip their arrows used in hunting. The fruit serves as rat poison.

Use

People use it like they use Casuarina, as a windbreak tree. It helps stabilize soil and control soil erosion. You can intercrop it with Acacia. The timber of this tree is valuable as people use it for flooring, boat building, furniture making, and carving.

Butterfly Weed - Asclepias tuberosa (Apocynaceae)

The butterfly weed has poison characteristics. It produces large amounts of nectar that attracts butterflies. Unlike other milkweeds, it doesn't have a sappy stem. It is the larval food plant of queen and monarch butterflies. Other names of this plant are Pleurisy root, Milkweed, Indian Paintbrush, Common butterfly weed, Chigger Flower, and Butterfly Milkweed. It belongs to the Apocynaceae family.

Description

This perennial plant grows 30 cm to 1 meter in height. It shows clusters of orange or yellow flowers that bloom in early summer to autumn. In the wild, these flowers exhibit severe reddish color. With its colorful umbels and lance-shaped leaves, spirally arranged 6-12 cm long and 2-3 cm broad. It has a spread of 1-1.5 feet. This herbaceous perennial will take 2-3 years to produce flowers.

Habitat and Distribution

They prefer dry sandy to gravelly soil. You can also see them growing on the margin of streams. It is native to many parts of the United States and Canada.

Toxicity

This genus contains toxic alkaloids, resinoids, and cardiac glycosides. Grazing animals usually avoid it. Large doses of this plant, if ingested, can cause diarrhea and vomiting in humans.

Use

It is a wonderful plant for a meadow garden. You can use it with other ornamental plants that attract butterflies.

Christmas Bush - Comocladia dodonaea (Anacardiaceae)

This is a species of tree belonging to the cashew family that has severe poison characteristics. You can find this small shrub along trails and open canopies. With its dark green leaves, it resembles holly. Other common names include Christmas bush and poison Ash. It belongs to the Anacardiaceae family.

Description

It has overlapping leaves 1-3 cm long with 3-5 spines midrib and laterally and flowers occur in threes in a red-purple color. Its fruits have a red color, 1 cm long.

Habitat and Distribution

It grows natively on the Caribbean Islands. You also find it on the Russian steppes and Grenada.

Toxicity

The urushiol poison is like that of poison ivy. This toxin is present on the surface of the leaves and in the latex. Irritation, swelling, burning, and itching skin are the symptoms of poisoning. The rash can last for several weeks.

Use

The extract of this plant helps to lighten the arms and faces. The timber is Bastard Brazil used for general purposes. If the resin contaminates the tools and clothing, dermatitis might result.

Milk Bush - Pencil Euphorbia (Euphorbiaceae)

This succulent shrub has branched cylindrical branches and is toxic if you touch its sap. They are drought resistant and tolerate light. It has many other names including Fire Stick Plant, Rubber Hedge Plant, Pencil Tree, Pencil Cactus, Naked Lady, Milk Bush, Finger Euphorbia, Sticks on Fire, Rubber Hedge Euphorbia, Petroleum Plant, Pencil Euphorbia, Pencil bush, Milk Hedge, Indian Tree Spurge, Finger Tree, and African Milk Bush. They belong to the Euphorbiaceae family.

Description

It has pencil-thick succulent toothed branches. It can grow up to 7 meters with finely striated, longitudinal twigs 7 mm growing in whorls. It has oval leaves 1-2.5 cm long and 3-4 mm wide which fall off fast. We see yellow flowers at branch ends and fruits are pale green with a pinkish tinge.

Habitat and Distribution

In Africa, they grow well in black clayey soil and grow natively in other places on the continent. You can also see it in tropical countries such as India, Philippines, Brazil, Vietnam, and Ghana.

Toxicity

The milky latex causes severe itching to the mucosa and skin. Exposure to the eyes can lead to blindness. If ingested, it causes burns to the tongue, lips, and mouth. Contact with the skin causes a burning sensation and redness.

Use

People grow it in fry areas to feed cattle. They may also use it for hedging.

East Coast Dune Flower - Helianthus debilis (Asteraceae)

This butterfly-attracting flowering plant has a hairy stem, and the heads follow the sun all through the day. The bright yellow color adds liveliness to any landscape. It has lots of common names such as East Coast Dune Flower, weak sunflower, beach sunflower, and cucumber-leaf sunflower. It belongs to the Asteraceae family.

Description

It is an annual or perennial herb that grows up to 2 meters in height. It doesn't tolerate the cold well. The leaves are alternate, variable in shape and size. The largest lead is 14 cm long and 13 cm wide. The inflorescence is a single flower head, or it might also be three or four heads. It has up to 30 phyllaries lance-shaped 1.7 cm long. It has about 20 florets of 2.3 cm long. The head is filled with florets that are white, yellow, or orange.

Habitat and Distribution

It is native to the United States and grows along the Gulf and Atlantic coast. In other parts of the world such as Cuba, Slovakia, Taiwan, Australia, and South Africa, it is an introduced species. It tolerates mildly salty environments, low-nutrient, and poor soils.

Toxicity

There is a rumor that the petals of this plant are toxic. But that is not true; they are nontoxic to dogs and cats.

Use

The flowers of this plant attract butterflies, and the fruits attract birds. People use it as a garden plant and a landscaping item.

Blackjack - Bidens tripartita (Asteraceae)

This species of plants spread far by spreading their seeds by sticking to the clothes. In some regions, there are not enough mammals, and here this plant evolved to spread by floating through the air. The seeds of this plant have barbs, and that is the reason for the many names it has.

It has lots of names such as blackjack, beggarticks, stickseeds, cobbler's pegs, tickseed sunflowers, Spanish needles, and burr marigold. Other names for this plant are leafy bracted beggarticks, three-part beggarticks, trifid bur-marigold, and three-lobe beggarticks. It belongs to the Asteraceae family.

Description
This herbaceous plant has triple leaves and grows up to 30-100 cm. It is an annual flowering plant with yellow flowers that collect at branch-ends in baskets of inflorescence. The fruit has hooks that attach to the clothing or the hair of the animal. Contains essential oils, vitamin C, carotene, mucus, and trace elements copper, chromium, and iron.

Habitat and Distribution
The plant occurs all over the world in warm temperate and tropical regions. You can find it in Africa, America, and Polynesia and some species also occur in Europe and Asia. It grows in meadows and swamps on the banks of lakes and rivers.

Toxicity
The plant might show some toxicity if ingested raw.

Use

This plant finds use as a honey plant in many households. It also serves as food for caterpillars such as the noctuid moth. It has diuretic and choleretic properties and is also used as a medicine for children affected by scrofula. It is digestive and appetite stimulant.

Adam and Eve - Lamium album (Lamiaceae)

This plant resembles the stinging nettle, but it doesn't sting, which is why we call it the dead nettle. People use Lamium album to make medicine. Local names for this flower include Adam and Eve in the bower, Helmet flower, bee nettle, and White archangel. It belongs to the Lamiaceae family.

Description

This herbaceous plant grows 50-100 cm in height with four-angled green stems. We have leaves that are 3-7.8 cm long and 2-5 cm wide. The shape is broad, triangular, with a serrated margin and the base is softly hairy. The petiole is 5 cm long while the flowers are in whorls of white. They grow on the upper part of the stem. The individual flowers are 1.3-2.8 long and are a favorite of bees.

Habitat and Distribution

We find it growing on the roadsides and hedges. It is a native to Eurasia extending from Japan in the East to Ireland in the West. We also see it in north Scotland and mainland Asia.

Toxicity

While it seems safe for most people to take by mouth, there is not enough evidence on the toxicity of the plant. We can get two phenylpropanoid glycosides along with quercetin and rutoside from the flowers.

Use

It is a favorite source of chlorophyll for some botanists. We also get plant pigment from this plant. In folk medicine, they use it to treat vaginal discharge and sore throat though there is no documented evidence for its efficacy.

Snapdragon - Linaria vulgaris (Plantaginaceae)

Other names for this plant are wild snapdragon, butter and eggs, rabbit flower, monkeyflower, lion's mouth, impudent lawyer, gallwort, gall weed, continental weed, calf's snout, bunnies mouth, bunny haycock, brideweed, yellow toadflax, Jacob's ladder, flaxweed, ramsted, and common toadflax. It belongs to the Plantaginaceae family.

Description

This perennial plant has spreading roots with erect to decumbent stems. The leaves are 2 -6 cm long, fine, threadlike, and blue-green in color. They are 1 -5 mm broad and the pale yellow flowers are 24-35 mm long and orange at its lower tip. They bloom in dense terminal racemes beginning midsummer until midautumn. Bumblebees visit these flowers. The fruit has the shape of a globose capsule 5 -11 mm long and 4-7 mm broad. It has plenty of seeds.

Habitat and Distribution

The plant grows along roadsides and dunes as well as disturbed and cultivated land. It grows natively in Britain and most of Europe, Greece, Norway, east and south of Pyrenees, and west Asia.

Toxicity

This plant is thought to be toxic if ingested in large amounts. But studies are yet to confirm this.

Use

People grow it for the cut flowers as they last for a remarkable amount of time in a vase. Squeezing the base of the corolla will make the plant "talk." In folk medicine, they make a tea out of the leaves of the plant as a laxative. It also helps cure jaundice and acts as a diuretic. Other ailments for which one may use it include piles, skin diseases, drowsiness linked to enteritis, and dropsy.

Cow Lily - Nuphar luteum (Nymphaeaceae)

This is a flowering aquatic plant that produces bright yellow flowers. Other names include brandy bottle, Wakas, SpatterDock, Cow Lily, Yellow Pond Lily, and yellow water lily and you will find it in wetlands and marshes. But it can also grow in water 5 meters deep.

The roots and leaves float on the water surface but it does not grow well in contaminated water. It belongs to the Nymphaeaceae family.

Description

This perennial aquatic plant has fleshy roots. The flower is solitary, terminal, hermaphrodite 2-4 cm diameter. It is held above the water surface. It has five to six large bright yellow sepals and many small yellow petals. It flowers from June to September. A bottle-shaped fruit with many seeds appears soon after flowering.

Habitat and Distribution

You can find it in ponds and lakes. Its native range is Eurasia, eastern United States, northern Africa, and West Indies.

Toxicity

There is a suggestion that the plant is poisonous, but there are no details. Resorcinol is the major component.

Use

The herbal tea helps cure many ailments, including cardiac problems, infertility, high blood pressure, and digestion. It is also used as a sedative, aphrodisiac, and emollient.

Oregano - Origanum vulgare (Lamiaceae)

People have used Oregano for a long time because of its flavor and its medicinal value. There are suggestions that it is toxic, but there is no evidence for this unless you eat a large amount. Oregano is a good combination of spicy foods. In folklore, people say that the goddess Aphrodite values it very much. People also call it wild marjoram and sweet marjoram. It belongs to the Lamiaceae family.

Description

This perennial herb grows 20-80 cm in height and has opposite leaves 1-4 cm long. The flowers are purple 3-4 mm long as spikes and are spade-shaped and purple. It does not survive the winter and prefers a hot, relatively dry climate.

Habitat and Distribution

It grows in Tunisia, Algeria, India, and Afghanistan natively. In other places around the world, it has become naturalized.

Toxicity

There are no suggestions that this plant is toxic. But if you ingest a large amount, it could lead to the burning of the mouth and swelling of the tongue.

Use

People use it to cook their food. People grow it to combat sadness and to bring luck.

Goldenrod - Solidago virgaurea (Asteraceae)

This flower is called goldenrods and there are 100-120 species in this family. This is a bright yellow flower and the state flower of Kentucky and Nebraska. The blooming of goldenrods in August is a reminder that school will soon begin, and children will need to get back. It belongs to the Asteraceae family.

Description

These perennials grow from rhizomes. Their stems are recumbent to ascending to erect. They can grow from 5 cm to over a meter in height. They are normally unbranched, but there might be branching in the upper parts of the plant. Both leaves and stems are hairless to pubescence of many kinds.

The basal leaves will get shed before flowering. Margins are entire but have heavy serration and a few leaves might show trivenation as preferred to pinnate venation. Flower heads are radiate type, but you might also see some discoid.

Habitat and Distribution

This open land species prefers drier soils. You will find it in the woodlands and savannah, roadsides, old fields, and grasslands. It grows in North America, northern Africa, Asia, and Europe, while medicinal variety originate in the countries in Eastern Europe such as Poland, Hungary, and Bulgaria.

Toxicity

Goldenrod may cause allergy in sensitive people. Goldenrod makes the body accumulate more sodium, thereby increasing blood pressure. People allergic to latex will experience the side effects of this plant.

Use

Though there is not enough evidence, people use it for gout, eczema, bladder stones, arthritis, hemorrhoids, kidney stones, wound healing, and inflammation of the throat and mouth. It also finds use to stop muscle spasms and as a diuretic.

True Comfrey - Symphytum officinale (Boraginaceae)

This plant with bright violet flowers grows by the riverbanks, lakes, and marshy regions. In Symphytum, there are 34 other species. This species is comfrey or common comfrey. Other names include true comfrey, slippery root consound, knitbone, boneset, cultivated comfrey, and Quaker comfrey.

Description

This perennial herb is a hardy plant that can grow 1-3 feet in height. It has a black, turnip-like root, and the leaves are broad and hairy. It has bell-shaped flowers of many colors. In most cases, they are purple or cream with stripes.

Habitat

It grows well in moist grasslands in many parts of Europe, such as the United Kingdom and Asia. It also grows in North America and Ireland. In the British Isles, the flowers tend to have a violet or blue color.

Toxicity

Long-term topical use of Boraginaceae is not advisable because it can lead to liver toxicity. Toxicity is due to the presence of the toxic compound pyrrolizidine alkaloid. These are also toxic to livestock.

Use

Boraginaceae has found a place in herbal medicine for more than 2,000 years. People use extracts, ointments, and pastes of the plant parts to treat joint disorders, bone fractures, hematomas, gout, and inflammatory disorders.

Bitter Buttons - Tanacetum vulgare (Asteraceae)

This plant germinates in spring, developing a single rosette. It has the common names common tansy, cow bitters, bitter buttons, garden tansy, and golden buttons and belongs to the Asteraceae family.

Description

This herbaceous flowering plant is perennial with finely divided compound leaves. The reddish and stout stem is erect and grows 50-160 cm and branches near the top. It has alternate leaves 10-15 cm long divided almost to the center in seven pairs, pinnately lobed. These are divided into smaller lobes having saw-toothed edges.

The flower heads are flat and button-like roundish and yellow. We see terminal clusters in summer and it has a smell like camphor or rosemary.

Habitat and Distribution

This herb is native to Asia and temperate Europe. It also grows in North America, South America, and parts of Oceania region. It grows on rough ground, meadows, grassy areas, riverbanks, and wastelands.

Toxicity

Some races contain a poisonous essential oil ketone beta thujone. Consuming these in large amounts will cause convulsions and damage the brain and liver. Others are safe to consume.

Use

People grow these attractive flowers in gardens. The oil of T. vulgare helps repel insects and flies. Leaf tips find use in the preparation of ointments and cosmetics. The unguent of the leaves is a cure for tumors in the tendons.

Yellow Foxglove - Digitalis grandiflora (Plantaginaceae)

This plant forms clumps in wooded regions with well-drained soils. It has spikes that help it self-seed. It has good medical value but dies out soon. The other names are large-flowered foxglove, yellow foxglove, and large yellow foxglove and it belongs to the Plantaginaceae family.

Description

The flowering stem will reach 70-120 cm in height. The leaves are glossy and green-veined. The yellow flowers are 3-4 cm long and spaced along the length of the stem. They have marked a mottled brown in the interior.

Habitat and Distribution

On mountains, it grows on warm slopes, stony places, and woodlands. It grows all over Europe and Western Asia.

Toxicity

This plant is toxic due to the presence of cardenolides.

Use

They use it for preparing a drug preparation that has cardiac glycosides, especially digoxin used for treating heart diseases.

Chapter Two

Mushrooms

Mushroom as a Food

Once considered an exotic item, mushrooms are fast gaining popularity as a superfood. Given its high protein content and fleshy texture, people the world over are including this food item into their daily menu. The reason is the rich culinary and nutritional value it has. The nutrient content includes minerals such as copper, ergothioneine, potassium, and selenium and B vitamins such as niacin, pantothenic acid, and riboflavin. But what happens when a person identifies a wild, poisonous mushroom as an edible one and cooks it?

Get Familiar with Mushrooms

Everyone is familiar with the pale, white spongy substance that we cook and eat. But what is it? And, how is it poisonous to us? The mushroom is a spore-bearing, fruiting body of a fungus. It is a growing thing like a plant. Since there is no chlorophyll (the substance plants use to make food), they must turn to some other way to make do. They break down dead plant material to derive their food source.

Types of Mushrooms

Mushrooms grow on the bark of trees and decomposing wood. They even grow on decomposing food. They prefer cool dark habitats that are humid. Before collecting them, it is advisable to identify them well to be sure that they are edible. So, what varieties are there in mushrooms?

1. Edible mushrooms

2. Poisonous mushrooms

3. Medicinal mushrooms

4. Psychedelic mushrooms

Proper cooking methods ensure complete safety. So, learn how to cook them first, and then you can find the key to perfect happiness with this exotic food item.

Edible Mushrooms

Do you know the best thing about mushrooms? They are cholesterol-free! They find their place as a food taste enhancer in gourmet cuisine across the world. We can harvest in the wild or cultivate these edible mushrooms. The ones easy to cultivate are available in the market. We can get those that are not always available such as morel, mutsutake, and truffle from private gardeners and collectors. It is important to note that one must prepare them with care, or they will be rendered toxic. Also, what is all right for some will not agree with others. Be sure to check for individual allergies before eating any mushroom. The best option is to choose and pick only very familiar ones that have been cultivated properly.

Poisonous Mushrooms

Toxic varieties of mushrooms contain poison that might be a gastrointestinal upset. You might also have diarrhea and vomiting. But there are a few that have deadly effects. Here are some of the poisonous mushroom varieties.

1. Alpha Amanitin Deadly

2. Phallotoxin nonlethal

3. Orellanine Deadly

4. Muscarine Potentially Deadly

5. Monomethylhydrazine Deadly

6. Coprine nonlethal

7. Ibotenic acid Potentially deadly

8. Muscimol Potentially deadly

9. Arabitol nonlethal

10. Bolesatine nonlethal

11. Ergotamine Deadly

Alpha Amanitin

It will cause fatal liver damage within three days of ingestion. This is due to the toxin death cap.

Phallotoxin

Many mushrooms have this toxin. It causes gastrointestinal upset.

Orellanine

This will cause kidney failure within three weeks of eating the mushroom. This is due to the presence of Redox cycler that is like paraquat.

Muscarine

Causes excessive salivation due to cholinergic crisis. We see this in many types of mushrooms. The antidote is atropine.

Monomethylhydrazine

Brain damage, seizures, hemolysis, and gastrointestinal upsets are manifested. The principal toxin in this metabolic poison is Gyromitra and the antidote is large amounts of pyridoxine hydrochloride given intravenously.

Coprine

When taken with alcohol, it causes illness.

Ibotenic acid

Excitotoxin is the primary poison.

Muscimol

Causes hallucinations and CNS depression.

Arabitol

Some people might have diarrhea.

Bolesatine

People get nausea, vomiting, and gastrointestinal irritation.

Ergotamine

The vascular system gets damaged. This might lead to a cardiac arrest or loss of the limb.

According to estimates, we know 100,000 species of fungi worldwide. About 100 of them are poisonous to us, but most of the poisonings are not fatal. Almost all the fatal poisoning is due to Amanita phalloides mushroom

Chapter Three

Poisonous Mushrooms

Death Cap - Amanita phalloides

This deadly poisonous basidiomycetes fungus is one of the many in the Amanita family. It grows natively in Europe with people introducing them when they plant their pine, oak, and chestnut.

The mushrooms appear in autumn and summer. They have greenish caps with gills and a white stripe. Cap color is variable and is often white.

The risk of poisoning increases because many edible species, such as Caesar's mushroom and straw mushroom, resemble this poisonous variety. What is more, the toxin present in these mushrooms, Amatoxins are thermostable. So, you cannot remove the toxicity by cooking them. The toxin present in half a death cap mushroom will kill an adult man. Historical deaths include those of Emperor Charles VI and Roman Emperor Cladius.

The main toxin is alpha amanitin. It causes liver and kidney damage leading to death almost always.

The fruiting body of the mushroom is large with a cap 5-12 cm in diameter, rounded to start with but flattening as it ages.

Difference Between Poisonous and Edible Mushrooms

There is no way to tell this difference. Only absolute experts can make out the individual mushrooms. Though there are general rules, one must not blindly use them because it could be dangerous. For instance, check out the following:

1. Mushrooms that grow on the wood are safe - How about the deadly Funeral Bell that grows on wood?

2. It is all right if you can peel the cap - What about Death Cap, the most poisonous of them all?

3. If animals eat them, they are safe - Not true! Many animals can gorge on the poisonous variety and not feel any difference.

As for beginners, use these rules to keep away from danger. If you do so, you might miss picking some lovely edible mushroom, but you will avoid a lot of danger.

1. If the mushroom has white gills, or a volva, or a skirt or ring around the stem, avoid it. The mushroom might be from the deadly Amanita group.

2. Avoid mushrooms that have a red cap or stem.

3. Don't eat anything that you have not identified well.

Also remember, the poisonous one will look like the harmless edible ones. In all, there are about 30 species of poisonous mushrooms ever found. Most of them belong to the same group. Here in this list, we discuss the deadliest ones (not discussed so far) and those that cause the most harm.

Fly Agaric - (Amanita muscaria)

This is the most popular toadstool one knows never to pick. It stands out because of its bright red cap and white spots. Children might occasionally try to pluck them if they found them growing nearby. But it is not understandable why dogs (and the occasional cat) will want to eat them, especially when it will kill them.

The main toxic agents in this mushroom are ibotenic acid and Muscimol. They act on the central nervous system and bring about coordination loss, nausea, and alternating states of sleep and agitation. You might even experience hallucinations. These symptoms begin to manifest one hour after ingestion; they are rarely fatal. The biggest effect is the exhibition of crazy behavior - our ancients did not miss out on that. They used the mushrooms for their rituals.

Wild and photogenic but toxic, Fly agaric grows near evergreen trees or deciduous trees like birch. The sprouting season begins in August and ends in December. The average height of the mushroom is 20 cm, and the width is 25 cm.

Cap: Hemispherical, it starts as red but might turn orange at times. It becomes flat and develops white to yellow scales. These are scales left behind from the volvic sack. If there is rain, the spots might get washed off so that the cap becomes smooth and spotless.

Gill: The gills remain crowded and white to cream in color. The gills do not join the stem.

Stem: It appears from a structure shaped like a bulbous sac (volvic sack), the whitish stem appears with the remains of the volvic sack sticking to it.

Skirt: This is higher up. It is pendulous and usually white, but the edges might be yellow.

Volva: We see shaggy, scaled rings around this bulbous structure.

Flesh: The flesh of the mushroom is white.

Toxicity

This mushroom is very toxic. The fatal dose is 15 caps. In the olden days, people used efficient detoxifying methods to render the mushroom edible. Reindeer lives this mushroom, and so people use the mushrooms to round up the reindeer. Sami and the people of Siberia used one local variety of this mushroom as an entheogen and intoxicant. An entheogen is a psychoactive substance that induces a change in mood, behavior, cognition, and perception. There is speculation about the use of this mushroom in places like North America, Scandinavia, Eurasia, and the Middle East.

The name comes from the practice of sprinkling powdered mushrooms in milk to kill flies. There is also the thought that the fly part refers to the delirium that people get from eating the mushroom.

Habitat

Fly agaric occurs natively in boreal and temperate regions of the Northern Hemisphere. They prefer the deciduous and coniferous woodlands and higher elevations of warmer regions. So, they grow in central America, the Mediterranean, and the Hind Kush regions.

Panther Cap - Amanita pantherina

This is an uncommon mushroom that has the names false blusher and panther cap. Contains a psychoactive substance, Muscimol. It belongs to the Amanitaceae family.

Description

Cap: The size of the cap ranges from 5 - 11 cm. It has a very finely striate margin, and the color is grey-brown or shiny brown. It remains domed at the start, but when the fruit matures, it tends to become flat. Remains of the universal white veil remain dotted on top evenly.

Gills: Free, white, and crowded; the gills are broad.

Stem: The stem of the Amanita pantherina is 6 - 12 cm tall, pure white with a hanging ring. It is quite chunky at the start but will turn thin and floppy as it matures.

Volva: This a little swollen stem base keeps the remains of the volva in the form of one or more wooly rings. Or, it might be as a helix with a narrow gutter below.

Spores: These are ovoid to in a broad way ellipsoidal and smooth. Size is 8-12 x 6.7 - 7.5 μm, it is inamyloid.

Odor/Taste: It has no distinct odor. When bruised, the mushroom smells like a radish. Warning: Do not attempt to taste this deadly toadstool.

Season: The season begins in August and ends in November.

Distribution
This is quite common in southern Europe. Southwest Portugal has plenty of false blushers.

Early Morels - Gyromitra esculenta

Gyromitra esculenta (family Discinaceae) is fatal if one eats it raw. But it is a delicacy in many places, including Eastern Europe, Scandinavia, and in North America, in the upper Great Lakes region. It has many names, including elephant ears, turban fungus, beefsteak mushroom, and brain mushroom because of its wrinkled and convoluted appearance. Other names are early morels and thimble morels. It looks like other delicious morels, and hence it belongs to the false morel group.

The esculenta species grows natively in Europe. It looks like Morels Morchella esculenta, which is delicious, and many people forage for it. The problem increases due to the name "esculenta," which means "good to eat." It caused many accidents in the countries of Eastern Europe, but with the advent of the internet, people have become conversant with toxic mushrooms.

Cap: Wrinkled and convoluted with brain-like deviations and folds. The color is pinkish-brown or reddish-brown and sometimes black depending on how old it is. It is partially hollow inside, and flesh is tan or whitish.

Gills: It doesn't have gills. Spores exist on the surface of the cap, and the underside of the cap is not visible.

Stem: Short and thin with deep vertical folds.

It doesn't have any distinct smell or taste.

Spores: These are smooth and ellipsoid. It has a red or orangish color.

Edibility: Toxic

Habitat: They form fruits on the ground, preferably under conifers. This fungus feeds on dead plant matter. Sometimes, they exhibit mycorrhizal nature (a symbiotic relationship where the plant supplies sugar to the fungus, and it supplies mineral nutrients and water to the plant) as well.

Distinguish a true morel from a false one.

The simplest way to check whether a morel is true or false is to slice it down the middle along the length. The stem of the true morel is hollow; the cap remains fused to the stem all along the length.

In the false morel, the stem is solid or filled with cottony tissue. The interior resembles a bell and clapper arrangement.

Benefits of Gyromitra esculenta

Often this mushroom dries out in the field, and if this happens, it will last for a long time. People claim that Gyromitra esculenta is edible. But there is no proof of this. Also, there is no proof of the medicinal value of this fungus. It contains a substance from which we can make rocket fuel.

Toxicity

There are many cases of death recorded due to the consumption of Gyromitra esculenta. But this doesn't deter people from eating it. The name of the toxin is Gyromitrin. In both humans and dogs, the toxin destroys the red blood cells. It also causes damage to the nervous system, the liver, and the gastrointestinal tract.

Symptoms appear within 12 hours of ingesting the toxic mushrooms. One can treat the patients, some survive, and many do not. Slight symptoms include diarrhea, headache, lethargy, and vomiting. Severe cases will result in coma, delirium, and death in five to eight days.

The lethal dose of gyromitrin is 19-49 mg per kilogram of body weight in adults and 9-29 mg/kg for children. This amounts to ½ to 1 kg and ¼ to ½ kg for children in one dose.

Treatment involves decontamination of the GI tract using activated charcoal, but one has to do this immediately after eating the mushroom. We treat severe diarrhea and vomiting by giving those affected intravenous fluids to rehydrate them. They check and correct the biological parameters as needed. One might need dialysis when the kidneys are failing. Also, we give blood transfusion if needed.

Cooking

When you cook the mushroom, the fumes arising from the food are toxic. Further, cooking doesn't destroy the toxins as they are thermally stable.

The toxin produced carcinogenic effects in mice; it might happen for humans also. Mushroom enthusiasts can eat this variety for years and not feel a thing and then get a bad batch one day, eat it and die. The degree of vulnerability also varies from one person to another. Pregnant women are more vulnerable.

Distribution

It is very localized in Ireland and Britain. Most often, they occur under pine trees. They introduced this throughout Europe and seen in parts of North America. They grow in sandy soil in temperate coniferous forests. You might see them growing under aspen trees also. The season is from April to July. For a few species, the sprouting may occur even as snow is falling.

A few people in Finland report burying fungus impregnated newspaper in the snow and harvesting them the following spring. It is abundant in the Cascade Range and the Sierra Nevada in northwestern North America. People have reported it from western Turkey.

Podostroma cornu-damae

This fungus looks frightening because it has misshapen appendages and bright red color. Japanese are courageous and a bit adventurous about their food. This fungus looks like Ganoderma lucidum, and people eat this poisonous mushroom by mistake. This belongs to the Hypocreaceae family.

Toxicity

Trichothecene mycotoxins are the main toxins in the fungus. These cause unpleasant effects and may result in death if left untreated. The side effects affect the brain, kidneys, and liver, among other organs. We see a depletion of the blood cells along with peeling the skin of the face and head, making it seem like that the person has radiation poisoning.

Range of Podostroma Cornu-damae

In the beginning, only Korea and Japan had this fungus. Today, we find it growing in northern Australia, Papua, and Java.

Deadly Webcap - Cortinarius rubellus

This mushroom is lethally poisonous and has the common name deadly webcap. It looks ordinary and resembles many edible species of mushrooms because it is tan to brown color. This C. rebellus genus is one of the seven in the Orellana group of the Cortinariaceae family.

Distribution

In the Northern Hemisphere, C. rebellus grows in parts of North America and parts of Britain and Wales. We also see it in northern European countries and Scandinavia.

Toxicity

The toxicity of C.rebellus is due to orellanine toxin, which destroys the liver and kidney if anyone eats the mushroom.

Description

This poisonous mushroom has an orange-red color and the smell of radishes. People mistake it for Cantharellus cibarius mushroom, which is very relished and edible. The initial symptoms of poisoning get delayed by 2-3 days, which is the same as the symptoms of flu. After one ingests the mushrooms, one has headaches and vomiting. Renal failure occurs if one does not get treated, and within a few days, the person will die.

If one undergoes dialysis along with other kidney and liver treatments, the person will survive.

Cap: The young mushroom has a convex orange-tawny brown cap that becomes flat on aging. But even then, it retains a slight umbo. It has remnants of the veil (cortina) attached to the underside of the cap. The surface is a little scaly and dry.

Gills: In young mushrooms, the gills are pale yellow. As the spores become mature, it turns rusty brown.

Stem: This is a little bowed instead of straight. It is paler than the cap but has the fibers mottled with red. This fibrous stem tapers a little toward the base. It is 5-17 mm across in size and 5-11 cm in height. The pattern is a little yellowish and snakeskin-like.

Spores: It is a rough subglobose to ellipsoidal. The print has a reddish-brown color.

Beechmast candlesnuff

The fungus growing from the outer casing of Beech seeds (called Beechmast) is the Beechmast Candlesnuff. They are insubstantial but tough and not edible. They belong to the Xylariaceae family.

Description

This is threadlike 2-5 cm long with a thickness of 0.5-1 mm. It might have branches. At the start black near the base, it becomes white at the tips. The entire fruitbody turns black eventually. The white tips are asexual spores. When the fruit ripens, the ascospores ripens within the flask like perithecia. On the outer surface of the fruit bodies, we have tiny bumps with minute holes that show the location of the perithecia.

Spores: Smooth and bean-shaped, 10-13 x 4-5.6 μm. The spore print is black.

Asci: Usually 120x6 μm. It has about eight spores per ascus.

Odor/Taste: Both are indistinctive.

Season: Throughout the year. Produces ascospores in autumn and the beginning of winter. During this time, the whole fruitbody turns black.

Distribution

It grows well in Britain and Ireland. We can also see it in many places in North America and Europe.

Toxicity

This fungus is not considered edible.

Angel Wings

This mushroom gets its name from the way it looks and its white appearance. People used to eat it as there were no reports of poisoning. But in 2004, 60 people in Japan became ill after eating angel wings, and in this group, 17 people died. In Japan, people call it sugihiratake. It belongs to the family, Marasmiaceae.

Description

This white-rot wood decay fungus favors conifer wood, such as hemlock. People ate angel wings in the past until they discovered it had toxic substances in 2004 in Japan. The toxin seemed to be an unstable amino acid called pleurocybellaziridine. The fruit bodies are white when they are young but become tinged with yellow as they age. The stipe is short or completely absent. The smell of the flesh is faint but pleasant.

Cap: The white, smooth, and stemless cap are 2-9 cm across with a split-sided incomplete funnel. It sometimes resembles a tongue with a lobed margin. The flesh is thin and white.

Gills: These are ivory white.

Spores: Smooth and globose, 5-6 μm in size.

Habitat: Well-rotted conifer timber with a covering of moss.

Season: In Ireland and Britain, it is from August to November.

Toxicity

When the toxin enters the body, the kidney becomes compromised. It is unable to filter it, and so the amino acid goes back into circulation in the body. When it reaches the brain, it breaks through the blood-brain barrier leading to a condition of encephalopathic condition. This is like that of motor neuron disease.

Distribution

We find it in northern England and Scotland. We see it in some places in North America and the cool regions of Asia. It also grows in northern mainland Europe.

Deadly Dapperling

This small mushroom is, as its name suggests, deadly. Many mushrooms in the Lepiota family to which this belongs, has the deadly amatoxin. This poison will destroy the liver and causes more than 70-80% of all mushroom deaths. The rate of death for untreated cases is 50% and 10% for those treated. The symptoms begin with gastrointestinal disturbances and then go on to cause renal failure. It has an equally deadly family member parasol. It belongs to the Agaricaceae family.

Distribution

We find it throughout North America and Europe. Lepiota Brunneoincarnata likes the grassy areas of fields and parks and it resembles the edible Fairy Ring Champignon (Marasmius oreades) and Grey Knight (Tricholoma terreum). We don't see it much in Ireland or Britain, but it is plentiful in the temperate parts of western Asia.

Description

Cap: This is 2.6-4.3 cm across at the start hemispherical that becomes in a broad way convex, so it is almost flat with a little umbo. The surface is felty and pinkish brown. This breaks into wooly scales to form irregular concentric rings that turn paler and more widely spaced toward the margins. The flesh is white. When it becomes mature, the diameter becomes 2.5-6 cm across.

Gills: These are creamy white, free, and crowded. The cheilocystidia are narrowly clavated or cylindrical.

Stem: The color of the stem is creamy white, and it has a pink flush. The length is 2.5-4.9 cm, with a diameter of 5.2-9 mm. The upper half is uniformly white, and the lower half dark-brown scaly with an indistinctive wooly ring.

Spores: These are smooth and ellipsoidal dextrinoid.

Odor and Taste: Faint, fruity odor, deadly taste - do not taste.

Season: In Ireland and Britain, we see it growing from July to November.

Toxicity

It is deadly poisonous, and so it is better to avoid this species of mushrooms. They contain deadly amounts of alpha-amanitin and can cause death if eaten.

Treatment

It is important to begin the liver protective measure immediately. Intravenous silibinin can reduce amanitin uptake. Other treatment options are n-acetylcysteine and penicillin G. You must also start rehydration for the patient.

Autumn Skullcap

This attractive mushroom may seem edible to many newbie mushroom enthusiasts. But they are deadly, and you will do well to avoid this fungus. They have the common name of Funeral Bell or

Autumn Skullcap, and their botanical name is Galerina marginata. They belong to the Hymenogastraceae family.

Description

These mushrooms have small brown caps that are sticky with a white annulus and rusty brown spore prints. Typically, they grow on rotted wood.

Cap: Conical to bell-shaped cap 1.5-2.5 cm diameter, sticky and brown when moist, yellowish when dry. The margin is striate when wet.

Pileus: This is typically glabrous with a cortina type veil in young fruits.

Gills: Attached to the stalk at the top, at the start yellow when young, they become brown when they develop.

Stalk: Tan or light brown, they are fibrillose below the annulus. They are hollow, and the base has dense white mycelium.

Annulus: This is white but becomes brown as the spores' number increases. It is on the top of the stalk, and the older mushrooms will not have this.

Spore print: Rusty brown.

Toxicity

The very toxic nature of this fungus makes it important to identify this genus well. Most of the mushroom hunters will look for the hallucinogenic Psilocybe mushrooms. Due to the superficial similarity of the Galerina marginata to the Psilocybe mushrooms, there is a real danger of picking poisonous ones. The easy way to identify them is to check the staining character. The Psilocybin Psilocybe fruits stain blue when bruised whereas Galerina does not. The Galerina might show a blackening that mushroom hunters misinterpret as a blue stain.

Destroying Angel

This is a deadly basidiomycete fungus of the Amanita family. They appear in autumn and summer and have the botanical name Amanita virosa. This causes many mushroom-related deaths. Everything in this fungus is white. When they are immature, they resemble many edible species, which leads to poisoning. The smaller varieties resemble the portobello mushrooms, and one mushroom is enough to kill an adult.

Description

The average height of the mushroom is 12 cm, and the diameter is 12 cm. It is an egg-shaped object covered by a universal veil. When it grows, the mushroom breaks free, making the edges of the veil ragged. At the start conical with in-turned edges, the cap becomes hemispherical and flat. The diameter is as much as 11.72 cm with a distinct boss. We can peel it and see the white flesh, but the center might be ivory in color.

The gills remain crowded and white like the stipe and volva; the stipe has a grooved hanging ring. It is 15 cm tall. The spore print is white, and the shape is like an egg, conical 7-10 μm in size. When we use iodine, it stains blue. The flesh is white and tastes like a radish. They turn yellow with sodium hydroxide.

Risk Factor

Since this mushroom resembles many edible varieties like Agaricus campestris and A. arvensis, it is not possible to make out the difference if the caps are not opened and the gills are not visible. In mushrooming, we regard the ability to peel as a sign of edibility, but this is a mistake. One should consider other factors also.

Distribution and Habitat

Amanita virosa prefers the beech tree but might grow in any mixed woodland. It prefers the mossy ground, and it grows in autumn and summer. We find it growing in the mountainous regions of Ireland and Britain. It also grows in the lowlands of Scotland and the conifer forests of Scandinavia. During July, August, and September, Destroying Angels appear in northern Europe.

Toxicity

The mushroom contains deadly doses of amatoxins. The initial symptoms of poisoning are vomiting, convulsions, delirium, cramps, and diarrhea within a day of eating the mushroom. Destruction of the kidney and the liver begins immediately, and one must begin the protective action; otherwise, the person will die.

Laughing Jack - Gymnopilus junonius

Gymnopilus junonius or Spectacular Rustgill grows on tree stumps, logs, and the base of trees. Orange in color, it is a colorful wood-rotting species that grows from spring to the beginning of winter. Some subspecies have the toxin gymnopilin. They have the names Laughing Jack or Laughing Jim (or Gymn) because Gymn means naked, referring to the bald cap, and Juno was the daughter of Saturn. She was also Jupiter's wife. It belongs to the Cortinariaceae family.

Description

The large, golden fruit looks tempting to eat. But it is not advisable since they are toxic.

Cap: The size of the cap ranges between 4-20 cm, large ones will be 30 cm across. It is convex at the start, and the margin is in-rolled, but it flattens out but retains a faint umbo. The golden appearance of the Spectacular Rustgill comes from the radial apricot or orange-colored fibers on a sienna or yellow background. The color of the flesh is straw-yellow to cream. It is quite firm and thick.

Gills: The gills of immature fruit bodies remain covered by a yellow cortina. It shrivels and appears as fragments along the rim of the cap and around the stipe. The crowded gills are adnate and attached in a broad way to the stipe. The color is buff to straw yellow that changes to a bright rusty color when the spores mature.

Stem: The color of the robust stipe is the same as that of the cap. The surface below the ring is fibrous. Spores gather, and soon the color turns rusty brown. The stipe is clavate (shaped like a club) or bulbous at the base. The stem is solid and has thick yellow flesh.

Spores: They are almond-shaped to ellipsoidal. They are 8 x 6 μm in size.

Odor/Taste: It has a faint fruity smell that increases when we cut it. They taste bitter.

Season: In Ireland and Britain, the season begins in June and ends in November.

Distribution

It grows natively in Australasia, Europe, and South America. It doesn't grow in North America. They like to grow on logs of conifer and hardwood. We find it growing profusely in the dense woodland near the river.

Toxicity

It contains hispidin and bis-noryangjin. They are like alpha pyrones we see in kava. Researchers have also found neurotoxins or oligiosoprenoids in the mushroom. It is best avoided as there is no evidence on the toxicity or edibility of the fungus.

Angel's Bonnet - Mycena arcangeliana

Names after the mushroom enthusiast, Giovanni Arcangeli, this species of mushrooms are small, and so we cannot use them for cooking purposes. Further, due to their strong odor, you cannot eat them. They grow on dead wood during autumn in singletons but prefer to form clusters. They have the common name Angel's bonnet or late season bonnet. It belongs to the Mycenaceae family.

Description

Identifying them because of the way they grow in tufts is not reliable because they also occur in singletons. But we can separate them from other members in the family through the coloring on their caps and gills.

Cap: They are 0.72-2.6 cm across (some can be 4.5 cm), conical that becomes bell-shaped. They are in a broad way umbonate smooth but having translucent striations. It has yellow to olive shades on top that become gray on drying and the color change is due to their hygrophanous nature. In the Angel's Bonnet mushroom, the cap becomes flat as they age.

Cheilocystidia: We see huge amounts of cheilocystidia standing out from the edges of the gills. They can grow up to 55 μm long. The pear-shaped cheilocystidia have many thin brush cells at their tips. The structure of the gill faces the same.

Gills: These are a little decurrent or adnate. Its color is white, which becomes pinkish gray with age. The edges are a little toothed.

Stem: It is 4.2-8 cm in height and 2-4 mm in diameter. At the apex, it is white. Young plants have a lilac tinge. The lower part of the stem is gray with tinges of olive. It doesn't have a ring; there are downy white hairs.

Spores: The spore print is white, and it is pip-shaped to ellipsoidal. It is amyloid and smooth 8x6 μm in size.

Odor/Taste: It smells like iodine. The taste is mild but indistinct.

Distribution

Mycena arcangeliana grows on dead deciduous wood in small groups, preferably ash and beech. It also grows on Japanese knotweed, bracken, and conifers. People report seeing it growing on the grassland. In the British Isles, it grows late summer to autumn. You can see it growing in many places in Europe.

Toxicity

The mushrooms are not edible. You can identify them by the smell of iodine.

Clouded Funnel - Clitocybe nebularis

This mushroom loves conifers. It has a foul odor, and you will find it growing alone or in groups. There are about 300 species of this mushroom in this genus. They grow on the ground, decomposing forest ground litter and belong to the Tricholomataceae family. People do not collect them for consumption.

Once they expand, the caps will grow to be 6-20 cm. The color will remain gray with a characteristic cloud pattern in the center. On the surface, it is common to find a felt-like bloom in the center and pale.

Gills: The white gills become cream with age, crowded and adnate, or a little decurrent toward the stem.

Stem: This has a swollen base and becomes 2-3 cm in diameter toward the top. The Clouded Funnel has a solid, smooth stem 6-12 cm in height. It is a little paler than the cap.

Spores: Smooth and ellipsoidal, they are 7.5x4 μm in size. The spore print is a pale buff to creamy white.

Odor/Taste: It has no distinctive taste. The smell is fruity.

Season: It begins in August and goes on to early December.

Distribution

They are common in Ireland and Britain. The Clouded Funnel also grows in Scandinavia and many parts of North America.

Toxicity

People used to eat it until recently it causes gastric upset often. The edibility depends on the place where it grows. People claim to have eaten the boiled mushroom without experiencing any discomfort.

Description

Clitocybe means sloping head. They grow in fairy rings, and the name Clouded Funnel refers to the cloud-like coloring on the cap.

Cap: Convex and becoming conical at the beginning, it will take up to a month for the cap to develop. Then, they flatten out and become a little funnel-shaped. The margins remain wavy, usually down-turned, and might be a little enrolled as well.

Waxy Laccaria - Laccaria laccata

This is a very variable mushroom that can look colorless and drab at times and vibrant and colorful in other places. The young ones have a red to orange color and grow well. Other names for this fungus are waxy laccaria and the deceiver and it belongs to the Hydnangiaceae family.

Description

This is a small, edible mushroom that enthusiasts consider a weed because of its plain stature. They have a mycorrhizal nature.

Cap: They are 2-7 cm across at the start of convex and become flat-topped when they become old. When it is wet, the cap of a young Deceiver has a reddish-brown or deep tan color but might sometimes be orange. When it dries out, the color becomes much paler and eventually turns white. Very old Deceivers might become distorted, and the cap will become funnel-shaped.

Gills: These are deep and broad and widely-spaced. They have short gills interspaced between them. The tan gills begin to lose color much before the cap turns to buff color. This happens because they get covered in spores.

Spores: Globose with spines, 7-10 μm with a height of 1.5μm. The spore print is white.

Odor/Taste: Both are indistinctive.

Season: The season begins in June and ends in November. In southern Europe, this gets delayed a little.

Odor/Taste: There is no distinctive taste or smell.

Distribution

We see the Deceivers in scattered troops in wooded areas and often in poor soils and it prefers cool weather and northern temperate zones. It develops a mycorrhizal relationship with many tree types, including Birch, Beech, and Pine. It grows widely in Europe, North America, Costa Rica, and Mexico.

Toxicity

These are edible and tasty, but one must watch out for small poisonous mushrooms.

Spotted Toughshank - Rhodocollybia maculate

Another name for Rhodocollybia maculata is the Spotted Toughshank. This is a tough mushroom, and so it is inedible. It grows in small groups in the litter and under conifers. It appears almost overnight with small reddish spots on its cap. These then erupt randomly over the gills. It belongs to the Marasmiaceae family.

Description

It is attractive, and fruits appear in groups. It doesn't mind where it grows.

Caps: They are 5-12 cm across obtuse conic to convex, eventually plano-convex flattening with a wavy margin. It turns upward to create an irregular saucer shape. It has a creamy pinkish white color with tan blotches or spots.

Gills: Adnate, adnexed, and uneven edges cream-colored crowded white. Develops brownish red, rust-like spots with age.

Stem: It is 5-11 cm in height 0.9-1.23 cm in diameter. It has no ring, is white and develops rust like spots in time.

Spores: Ellipsoidal to subspherical and smooth, 6x5 μm. The spore print is cream with a pinkish tinge.

Odor/Taste: No distinctive odor, a taste is so bitter the fungus is inedible.

Season: It begins in June and lasts until November.

Distribution

It is saprobic, meaning it lives in a relatively oxygen-free environment that is rich in organic matter. You find it on needle litter or well-rotted wood underneath conifers. It grows in central and northern countries in Europe. You also see it in many places in North America and Canada.

Toxicity

Though it is not toxic, the taste is so bitter that people do not collect it for consumption.

Verpa bohemica

This mushroom early in spring, which is why many people mistakenly call it Early Morel. But it continues fruiting through the true morel period. It has a resemblance to the half-morels Morchella punctipes and Morchella populiphilia. But half morels are only half free. When you compare it to Verpa bohemica, we see that it hangs completely free.

Description

Another way to separate V. bohemica from the half-morels is to slice them along their length. The half-free morels are hollow, but V. bohemica has a cotton-candy type of flesh inside.

Cap: The pale yellow or brown cap is conical or bell-shaped 2-4 cm across and 2-5 cm in height. The surface has wrinkles or brain-like contusions. The ridges are darker than the pits, dry or moist, and bald or fuzzy. The color is tan to brown, and the underside is white.

Stem: It is 8.5-24.3 cm long and 1.5-3 cm across, a little tapered to the top or bottom dull yellow to creamy white. It stains orangish when bruised.

Flesh: Both the stem and cap flesh is thin and hollow. Interiors have wispy white fibers.

Odor and Taste: There is no distinctive odor or taste.

Spores: The size is 48-85 x 14-24 μm. They are smooth, thick-walled, and elongated ellipsoid.

Distribution

We find V. bohemica growing in Asia, Europe, and North America. It grows on the wood on the ground after the snow melts.

Toxicity

It is suspect and may cause gastrointestinal discomfort in sensitive people. These people will experience dizziness, diarrhea, abdominal pain, muscle cramps, fatigue, and bloating.

Poison Pax - Paxillus involutus

This mushroom is a varied species and grows under many trees. This basidiomycete fungus is widely distributed and stains brown when bruised. This deadly gilled mushroom has the shape of a funnel. When bruised, they stain brown. Its other names are poison pax, common roll-rim, and brown roll-rim. It belongs to the Paxillaceae family.

Description

It grows in urban areas and the woods alone, gregariously, or scattered in summer or fall. It is widely distributed in North America.

Cap: It is 4.6 -14.3 cm convex to in a general way convex strongly in-rolled margin is cottony. It becomes centrally depressed and plano-convex dry or sticky. It is finely hairy or smooth, olive-brown, brown, or yellow-brown.

Gills: It runs down the stem and is separable as a layer. Close or crowded, it becomes pore-like or convoluted near the stem. The color is pale, olive, pale cinnamon, or yellowish. It bruises reddish-brown to brown.

Stem: The length is 2-9 cm and 2.1 cm thick. It tapers toward the base and dries smooth or finely hairy. It has the same color as the cap.

Flesh: It is yellowish, thick, and firm. It discolors to brown when exposed.

Odor and Taste: Not distinctive or acidic in taste. The odor is not distinctive or slight fragrance.

Spore: The prints are yellowish-brown to purplish-brown.

Distribution

There are no specialized nutrient needs; the mushroom has a broad range of hosts. The P. involutus makes antifungal compounds that protect the host plant against root rot. very abundant, the fungus grows everywhere in Asia and Europe. You can see it in India, China, Turkey, Iran, and Japan. It is widely distributed across North America and southwestern Greenland.

Toxicity

The mushroom has an antigen that triggers the immune system to attack the red blood cells. Complications include acute respiratory failure, shock, disseminated intravascular, coagulation, and acute kidney injury.

Wooly Gomphus - Turbinellus floccosus

This mushroom has the names scaly, wooly, or shaggy chanterelle. And until 2011, it went by the name of Gomphus floccosus. Another name it has is wooly gomphus. It is a cantharelloid mushroom belonging to the Gomphaceae family.

Description

The mushroom is mycorrhizal conifers such as fir, spruce, pine, and hemlock. The fruiting bodies are orange and vase or trumpet-shaped.

Cap: The orange fruiting body is 5-15 cm across at the start cylindrical with the center sunk in. As it matures, it becomes deeply depressed. The shape might be infundibuliform or cyathiformis. The surface is dry, covered with erect and recurved but small scales. The color of the scales is buff, yellow to pale orange. The flesh is white and fibrous, and margins undulate.

Hymenium: Narrow, blunt edges, and wrinkles that give rise to irregular, anastomose veins cream to yellow.

Stem: This is not distinct from the hymenium. It is 8-20 cm in height and 1-3 cm at the base. It is buff-colored.

Odor and taste: It has mild odor and taste.

Spore: Entire, minutely wrinkled, ellipsoidal, and thick-walled. It is 11.5-14.5 x 7-8 µm.

Edibility: Some people can eat it and not feel anything. But it is advisable to avoid eating it.

Distribution and Habitat

The fungus forms a symbiotic relationship with many conifers, including fir, Douglas fir, Pine, European Silver fir, and momi fir. The Gomphus floccosus grows in the North American coniferous forests. We see T.floccosus more abundant in the older stands of the forest where decomposed wood is present.

Toxicity

It contains the toxin Norcaperatic acid. This toxin is a gastrointestinal irritant and can cause vomiting, diarrhea, and stomach pain if you eat the mushroom.

Chicken of the Woods - Laetiporus sulphureus

This mushroom has an appetizing common name, Chicken of the Woods. The group of mushrooms in the genus Laetiporus is small. They are a small group of soft fleshed polypores with a bright orange or yellow color. These "Chicken of the Woods" mushrooms cause brown wood rot in both the hardwoods and conifers and belong to the Fomitopsidaceae family.

Description

With age, the flesh becomes tough and inedible. They collapse and decay into black mush.

Cap: The brackets are soft and spongy with wavy edges and broad margins. When they grow older, they become thin and pale. They range in width from 10-40 cm and have a

thickness of 3-12 cm. Color lies between egg yellow and creamy yellow with bands of an orange and pink tinge. When it is moist, the flesh is yellow-orange.

Tubes and Pores: There are tiny oval or round tubes on the underside of the brackets. These are 2-3 per mm with a depth of 15-30 mm. They have a very pale yellow or white color.

Spores: in a broad way, ovoid to ellipsoidal 5-7 x 3.5 μm. The spore print is white.

Odor/Taste: The smell is mushroomy; it tastes a bit sour.

Season: It grows through the summer and autumn periods.

Distribution and Habitat

It has a saprobic nature feeding off dead hardwood timber, especially sweet chestnut, oak, and beech. At times, we see it growing on willow and cherry trees. If you are lucky, you might spot this fungus growing on conifers such as yew.

We can find it growing from Alaska to California in North America. It is also abundant throughout Europe. In the Mediterranean region, we find it growing on Eucalyptus and Ceratonia.

Toxicity

This fungus is edible. The poisoning causes only mild symptoms such as vomiting, diarrhea, nausea, and cramps in some people.

Common Roll Rim - Paxillus involutus

The shape of this mushroom might suggest that it belongs to the funnel mushroom group. Here the color of the gills is white instead of brown like those of the funnel mushrooms. The spores are brown, unlike the funnel mushrooms that have white spores. The common names of this fungus are Common Roll Rim, Poison Pax, and Brown Roll Rim. This basidiomycete mushroom belongs to the Paxillaceae family.

Description

The Brown Rim Roll form ectomycorrhizal relationships with conifers and hardwoods. Studies show that the P.involutus might be a complex species rather than an individual species.

Cap: The ochre cap is convex at the start and becomes centrally depressed. There is a noticeable umbo, and the color becomes brown gradually toward the edges. The edges roll under to the gills, and this is the reason for its name. The edges are usually fluted. When the weather is dry, it is downy; it becomes viscid when it gets wet. Growth of the cap is between 5-12 cm, and the downy surface becomes slick when the mushroom ages.

Gills: These are pale ochre at first but become brown when the fungus becomes old. Rusty spots appear, and the places where the gills bruise will turn rusty brown. The crowded gills remain deeply decurrent.

Stem: This is 7-13 mm in diameter and 6-13 cm in height. They remain bent and grow parallel to the host plant. They start as light ochre but become chestnut brown as they grow old or get bruised.

Spores: These are smooth, ellipsoidal, and 7.4 x 8.2 and 5.2-6.2 μm. The spore print is sienna.

Distribution

They like to grow under birch and other broad-leaved trees; they prefer acidic soils. They grow in many parts of the world, including Britain, Ireland, New Zealand, Australia, North America, and Asia.

Toxicity

Brown Roll Rim is toxic; this is now established. The toxin causes gastric upsets and, in extreme cases, death. This happens because of an antigen that attacks the red blood cells through the immune system. This causes respiratory failure, kidney injury, and shock.

Scarletina Bolete - Neoboletus luridiformis

This fungus of the bolete family produces pores and tubes beneath their caps. Known as Boletus luridiformis before and wrongly as Boletus erythropus, this fungus has many names including scarletina bolete, dotted stemmed bolete, dotted stem bolete, and red foot bolete. Eating raw red foot boletes causes gastric upsets. This fungus belongs to the Boletaceae family.

Description

This large solid fungus has a convex to hemispherical cap. It is bay brown and up to 20 cm wide. They are quite felty when they are young. Its spores are orange-red and small and become rusty when they age. They turn black to blue when they bruise. The tubes are yellowish-green, which becomes blue when they bruise. The stem is colorful, densely dotted with red and gross 4-11 cm tall. It doesn't have any network patterns. The flesh stains blue when cut or bruised. It has no significant smell. The spore dust is a greenish-brown olive color.

Distribution and Habitat

We find it growing at all places in Europe in the summer and autumn seasons. It associates with spruce and beech. They prefer acidic soil over everything else.

Toxicity

Though this mushroom is edible, it is often confused with The Devil's bolete Rubroboletes satanas. Upon cooking, they turn black, and people may not like a dark dish.

Trametes suaveolens

Many of the Trametes species are of industrial and environmental interest because of their link with enzymes. The genus contains about 50 species and grows worldwide. They have a characteristic pileate basidiocarp and hymenium without cystidia. It has green tinges of algae and grows on dead and living broad-leaved trees, particularly willow and poplar. It belongs to the Polyporaceae family.

The bracket fungus forms in many layers, especially when it grows on dead stumps.

Description

This bracket is very light or creamy white throughout. The upper surface is undulating and downy, algae are growing most of the time. The brackets are 6-12 cm across and are round, to begin with, but the edges get sharper as they mature. Its upper surface is fertile, while the lower one is infertile. The flesh is tough, and there is no stem.

Tubes and pores: It has 10-15 mm white tubes that end in a little elongated or round buff, white, or pale-gray yellowish pores. They remain spaced at 0.5-1 mm and appear more stretched out when the pore surface slopes. Upon bruising, it turns brown.

Spores: Allantoid (meaning sausage-shaped) smooth 8-1.3 x 4-4.3 µm, inamyloid. The spore print is white.

Smell/Taste: Fresh mushrooms have a strong smell of aniseed; there is no significant taste.

Season: All-around the year, sheds spores in autumn.

Distribution
It grows in temperate regions of mainland Europe and can also be seen in parts of Asia and in many places in North America.

Toxicity
Although no one has reported it as toxic, the mushroom is too tough to eat.

Typhula fistulosa

The simple unbranched club-shaped stems that look like erect worms, Typhula fistulosa are flexible and soft and they sway even in the gentlest of breezes. They are inedible and their names mean a tube or pipe. They belong to the Typhulaceae family.

Description

Fruitbody: They remain flattened laterally a little and have longitudinal grooving. In a few instances, they are straight, but often they are wavy. They are simple clubs with mitriform or rounded tips. The colors range from light brown to ochre and sometimes orange when the tips are more fertile. Each stem is 5-30 cm in length and about 10 mm in thickness.

Spores: These are smooth and teardrop-shaped. Size is 10-17 x 5.4-8.6 μm, inamyloid, hyaline.

Distribution and Habitat

We find them in deciduous woodlands saprobic in leaf litter on the ground. At times, they grow from rotting wood, especially birch. We find it in most places in Ireland, Britain, and mainland Europe. We also see it grow in parts of North America.

Toxicity

Pipe Club fungi are inedible.

False Chanterelle - Hygrophoropsis aurantiaca

Known as False Chanterelle, Hygrophoropsis aurantiaca grows in heathland and woodland in many parts of the world. We sometimes see it on woodchips we use in gardening. People mistake it for the much-prized edible Chanterelle Cantharellus cibarius. The False Chanterelle belongs to the Hygrphoropsidaceae family.

Description

Pileus: It is 2.5-7 convex and broad and almost plain when mature. Often the disc remains depressed, margin incurved, and becoming decurved. The surface is dry and finely tomentose. Color ranges from yellow-brown to orange-brown or orange. It is darkest at the disc but sometimes arranged in concentric bands. Color fades as it ages, the flesh is thin, pallid to pale orange.

Lamellae: They are close, narrow, repeatedly forked, decurrent, and orange usually brighter than the cap.

Stipe: It is 2-7 cm tall, 0.5-1 cm thick equal or enlarged at base, central or eccentric in the attachment. It has a dry surface concolorous with the cap; there is no veil.

Spores: They are 5-7.5 x 3-4.2 μm smooth elliptical, dextrinoid. The spore print is white.

Edibility: There isn't enough evidence on whether they are safe or toxic. Some claim it is safe, but others say it is toxic.

Season: The season begins in August and goes on until November.

Distribution

Though it is a sparse genus with five species, we see it grow in Ireland and Britain. It is well-distributed across mainland Europe. North America has its share of this mushroom as well.

Bitter Poisonpie - Hebeloma sinapizans

This mushroom has a strong smell of radish along with a bulbous stem base. It has the common names rough stalked hebeloma and Bitter Poisonpie and belongs to the Hymenogastraceae family.

Description

Cap: The color is yellowish-ochre to pale buff and sometimes becoming light brown with a cinnamon tinge. It has a bell shape along with an involute, meaning a little in-rolled margin. This remains incurved until the cap expands. It becomes in a broad way concave or a little umbonate and becomes flat when it ages. It is sticky when wet, smooth, and silky when dry. In most cases, the margin is a little wavy and even lobed at times. It measures 5-13 cm in diameter.

Gills: It has a clay buff color that becomes reddish-brown with age. It is emarginate and notched, crowded. It doesn't exude watery droplets that dry and become brown spots.

Stem: This is very pale yellow or white. It becomes mealy toward the apex and scaly below. The base is swollen, entirely cylindrical. Length is 5-12 cm, diameter is 1-2 cm. There is no ring.

Spores: Lemon to almond shape covered with surface warts. Size is 10-13 x 6.5-8.3 μm, slowly dextrinose.

Odor/Taste: It has the smell of radish along with a bitter taste.

Season: It begins in July and ends by November.

Distribution

It grows in tufted groups in broadleaf woodlands and forms association with beech trees. We see it grow in all places in mainland Europe where it pairs mycorrhizal with both beech and oak trees. In North America, there is a species with the same name.

Toxicity

These mushrooms are poisonous. The name, Bitter Poisonpie is enough to warn people about the nature of this mushroom. This is not meant for eating purposes.

Lilac Bellcap - Mycena pura

Mycena pura, the Lilac Bellcap or the lilac bonnet, is very colorful. This color depends on the habitat and the amount of light that falls on it through the tree canopy. In grasslands, it appears in its yellow forms, and it becomes difficult to identify.

Description

When it is young, the colors lilac or purple almost always are present. Other hues might also be present, depending on where it is growing.

Cap: It is 2-6 cm bell-shaped or convex becoming flattened. The margin has lines and bald. It can occur in both dry and moist states. Typically, lilac or purple when young, they become pale and develop other shades that include pinkish-brown or red, yellow, or white.

Gills: These have a tooth to attach to the stem. They may be close or almost distant. The color is white or a little purple to pinkish. It develops veins when it matures.

Stem: This is 4-10 cm long and 2-6 cm thick. They are hollow, equal, and smooth but might sometimes have tiny hairs. The color is white or flushed with the cap color.

Flesh: It is not significant, watery gray to white.

Odor and Taste: Odor radish-like at times, it may not be present. Taste is strongly radish-like.

Spores: Subcylindrical to ellipsoidal and smooth. Size is 6-9 x 3-4 μm amyloid. The spore print is white.

Distribution

We see it growing all over Ireland and Britain. It also grows at all places in mainland Europe and North America.

Toxicity

Mycena pura contains the toxin muscarine, which is deadly, but because it is present in low concentrations, it doesn't cause death immediately. Poisoning symptoms include vomiting, diarrhea, cramps, and dizziness.

Entoloma serrulatum

Seeing mushrooms in blue-black color is rare. But there are many blue specimens of which Entoloma is perhaps the loveliest.

Description

Cap: The cap is 1.5-3.5 cm in diameter. It is convex at the start and expanding; it becomes umbilicate or in a broad way convex. The young ones are blue-black while the older ones are brown. They are radially fibrillose; the texture is finely scaly or silky.

Gills: It remains adnexed and sometimes has a decurrent tooth. They are broad, and the spacing is moderate. It has jagged or serrated edges with a pale bluish-white color. This turns flesh pink and has blue-black edges.

Stem: This is 4-7 cm long and 2-3 mm thick. It is silky with blackish dots near the apex, blue-black, and smooth below. Also, it is hollow and cylindrical with no ring.

Spores: The spores are irregular, and the side view shows 5 to 7 angles. Size is 9-12 x 6.5-8 µm. The spore print is pink.

Odor and Taste: It has a little mealy smell. The taste is not distinctive.

Season: Fruiting begins in early summer. It goes on until late autumn when the weather is mild.

Distribution

It grows gregariously in woods in moist regions. At times, it fruits from moss-covered woods. You can find it widely distributed in North America, especially in Arizona. It occurs throughout Europe and in the Black Sea region of Asia. They grow under deciduous woodland such as beech, oak, and occasionally birch.

Toxicity

This fungus causes 11% of the deaths due to mushroom poisoning in Europe. It is gastrointestinal with symptoms including vomiting, diarrhea, and headache that occur within half-an-hour to two hours after ingestion. The symptoms can last for two days. You might also see psychiatric symptoms like mood disturbances and acute liver toxicity along with delirium. At times, the symptoms may last for months.

Alcohol Inky - Coprinus comatus

This common fungus grows on lawns and alongside gravel roads. Other names for this shaggy mane, lawyer's wig, and shaggy ink cap. The technical name is Coprinus comatus. It belongs to the Agaricaceae family.

Description

Cap: It is smooth and egg-shaped and has a small area in the center covered with flattened scales. There are no veil fragments. The cap later becomes bell-shaped. There is a slight umbo, and then it deliquesces to the margin. Its color is grey-brown or gray, and then it turns black. The caps are 3-7 cm across.

Gills: The Common Inkcap has free, crowded gills that are white at first, then turns brown and after a while black. It then auto digests.

Stem: It is white with reddish-brown fibrils near the base. The thickness is 7-15 mm, and the height is 5-12 cm.

Spores: Almond shape to ellipsoidal 7.5-12 x 4.5-6 μm. It has an apical germ pore.

Season: It begins in May and goes on to November.

Odor and Taste: It is not distinctive. Warning: It is poisonous if consumed within a day or two or having alcohol.

Distribution

It grows in unexpected areas such as the green regions in towns and we see it in meadows and grasslands of North America and Europe. It is an introduced species in Iceland, New Zealand, and Australia.

Toxicity

This fungus is toxic if you eat it along with alcohol. If you do this, the effects are severe. If you plan to take or have taken alcohol within three days, then do not eat this mushroom.

Shaggy Mane - Coprinus comatus

We can see this fungus grow on our lawns and along the gravel sidewalks. It appears as a small white cylindrical growth and is easy to spot because it grows in bunches, spotting the area with white. Other names for this fungus are shaggy mane, lawyer's wig, and shaggy ink cap. The technical name is Coprinus comatus, and it belongs to the Agaricaceae family.

Description

Cap: The bullet-shaped cylindrical cap is shaggy, scaly, and white. They are 2.5-5 cm wide, 5-15 cm in height, and gooey. Scales remain upturned; color ranges from white tan to reddish-brown.

Gills: These are free, close, and white. It goes on to become black and inky.

Stem: It is hollow and fibrous, white to tannish. The stem has a partial veil on its middle to the lower area.

Flesh: The flesh is soft, white, and broken with ease.

Spores: These are black. It lacks pleurocystidia. The spores measure 10-13 x 6.5-8 μm.

Odor and Taste: Taste is mild, it produces a large amount of liquid when we cook it. One must not confuse it with the common Inkcap, which is toxic. This is due to the presence of coprine toxin in the common Ink Cap. Symptoms of poisoning include palpitations, diarrhea, vomiting, and a metallic taste in the mouth.

Season: Starts in June and ends in November.

Distribution

We can see it growing in common areas in the urban regions. We can see it growing freely in Australia, Europe, and parts of North America.

Toxicity

You can eat the fungus before the gills turn black. Beware of the common Inkcap, which is poisonous.

Sulfur Tuft

This common woodland mushroom grows in places where no other mushroom will. It is a saprophagous fungus with small gills that grows in large clumps on rotting trunks and dead roots of broadleaved trees.

Description

Cap: It is sulfur-yellow with tan toward the center of the cap. It is concave or a little umbonate and the cap margin has dark velar remnants attached to it. The cap is 2-7 cm across in size. The flesh of the cap is firm and yellow.

Gills: These remain crowded and adnate. The color of the Sulfur Tuft gills is sulfur yellow, becoming olive green and blackening progressively with the ripening of the spores.

Stem: The stems are almost concolorous with the cap. It becomes brown toward the base. The thickness is 5-10 cm, usually curved, and has a length of 4-13 cm.

Spores: They are smooth and ellipsoidal 6.7-84 x 4-4.4 µm. It has a small germ spore. The spore print is purplish-brown.

Odor and Taste: The taste is bitter, but the odor is indistinctive.

Season: This fungus grows all year-round.

Distribution

This grows well on the deadwood of coniferous and deciduous trees. It is abundant in northern Europe and North America.

Toxicity

The Sulfur Tuft mushroom is toxic due to the presence of steroid depsipeptides. In humans, the symptoms of poisoning will manifest after 10 hours. The first signs are vomiting, nausea, diarrhea, collapse, and proteinuria. A few cases have shown impaired vision and paralysis.

Blusher - Amanita rubescens

The common name for this fungus is Blusher and is one of the many species in the Amanita genus. This is an edible and tasty mushroom much sought after by culinary enthusiasts. What makes it recognizable is the pink color near the stem bottom. They get their name because they blush when you bruise or cut them. It belongs to the Amanitaceae family.

Description

Cap: The cap color doesn't give any idea of the identity. The color varies from white to different shades of pink and brown and even black color. It is 5-19 cm in diameter, depending on how much it has expanded. Color is brownish-pink and variable. It has irregularly distributed gray and off-white fragments of the universal veil. It has some shape at the start but flattens when maturing. At times, it becomes funnel-shaped. The cap and gills, when damaged, turn dull red or deep pink. When it is wet, the veil fragments get washed away from the cap.

Gills: These are white and adnate almost free of the stem. They remain crowded. In the mature Blushers, the gills have pink or red spots. Also, if you handle them, the mushroom blush pink or deep red.

Stem: The stem of the blusher is 7-14 cm in length, and the diameter is 1-2 cm. Above the ring, the stem has a reddish-brown color; there might be deep pink flecks. Below this, the stem bruises pink. When it ages, the stem becomes hollow, and it bears a hanging ring that is fragile and thin. It remains grooved and often ragged. Immature specimens will show the volva. When the mushroom becomes mature, it will disappear, and the stem will become swollen.

Spores: Ovoid or in a broad way ellipsoidal, they are smooth and 8-9 x 4.7-5.8 µm, amyloid.

Taste and Odor: Both the taste and odor are not distinctive. When they get infested by maggots, they can smell unpleasant.

Season: The season begins in June and ends in October.

Distribution

The Blusher grows profusely everywhere in Ireland and Britain. It is also found in North America and mainland Europe. We also see the Blusher in South Africa, where it is an introduced species.

Toxicity

The Blusher is edible, but one has to cook it well. The hemolytic toxin can cause anemia if you eat the mushroom raw and is the reason people don't choose this type of mushroom for their diet. The extreme variability makes it difficult to identify. Some are only 2.5 cm across, while a few others are as much as 20 cm across. Also, the cap color is variable. Some are brown, while others are bright metallic silver.

St. George's Mushroom - Calocybe gambosa

This is an edible mushroom that grows grass verges, roadsides, and fields. It has the common name of St. George's mushroom because it appears in the UK on St. George's Day (April 23). In Italy, a warmer country, it appears in March. People consider it a delicacy when fried in butter. It belongs to the Lyophyllaceae family.

Description

Cap: It is 5-14 cm in size, at the start almost spherical and becoming convex. At times, it is almost flat. The cap of the St. George's Mushroom is, at times, misshapen. In most cases, it retains a margin that is a little incurved. It has a white and smooth surface with a light brown tinge. This will become tan as it ages. The flesh is firm and white and susceptible to infestation by maggots. So, it is better to collect only fresh, young mushrooms.

Gills: The mushroom has sinuate gills. They remain crowded, narrow, and white.

Stem: It is 2-4 cm wide. Solid with a curve and thicker at the base, the length is 3-7 cm in height. It doesn't have a ring.

Spores: Smooth and ellipsoidal.

Odor and Taste: it has a floury taste and a mealy odor when it is raw. Cooking removes the flavor and smell and makes the mushroom delicious.

Season: It begins in April and ends in June. In the southern European countries, it begins a month earlier. In countries like Scandinavia, the beginning could be as late as June or July.

Distribution

We can see St. George's Mushroom in cropped pastures, but it will not be near deciduous trees. It is very common in lime and chalk rich regions. We see this in North America and throughout Europe. In Germany, this mushroom appears after May. In the warmer climate of Mediterranean countries, it will begin to grow in March.

Toxicity

It is safe to eat even uncooked. The mature mushrooms tend to get infested by maggots and so it is better to use only young mushrooms.

Brick-Red Tear Mushroom - Inocybe erubescens

This mushroom is one of the many in this genus to cause death; it grows in small groups in the litter and is associated with beech. This fungus has many names such as Brick-Red Tear Mushroom, Deadly Fibrecap, Red-Staining Inocybe, and I. patouillardii. It belongs to the Inocybaceae family.

Description

Cap: The cap is 3 - 7 cm across conical at the start and flattens as it matures. It keeps the pointed umbo with radial fibers that streak out and slowly redden from the enrolled margins. The surface of the cap bruises red. During the dry spells, the caps become marginally lobed and tend to split radially from the edge.

Gills: Adnexed or adnate and crowded, the gills are white, to begin with, and slowly turn pinkish brown inwards from the margin. On bruising, the gills also turn red. Cheilocystidia: These are the cystidia on the edges of the gills. They are cylindrical and thin-walled. Also, a little clavate.

Stem: The stem is white, 8-10 cm tall and 1-2 cm diameter with longitudinal fibrils. These stain red. We see a slight swelling at the base of the stem. If bruised, the flesh of the stem doesn't change color.

Spores: They are smooth and bean-shaped. Size is 10-13 x 5-6.7 μm. The spore print is dull brown.

Odor and Taste: The young and fresh mushrooms have no odor, but the older ones have an unpleasant smell. The taste is mild. Warning: This is a deadly mushroom, and so do not attempt to taste it.

Season: The season begins in late June and lasts until September.

Distribution and Habitat

These grow best in chalky soils and beech woods, growing on leaf litter during the summer and spring. It grows in Turkey and southern Europe.

Toxicity

This is a poisonous basidiomycete that causes nausea, vomiting, flushing, abdominal pain, and diarrhea. It is best to leave this mushroom alone.

Chapter Four

Medicinal Mushrooms

Medicinal Mushrooms

Improve clarity of the mind, get better sleep, and spice up your sex life with mushrooms. You must know how to identify the right mushrooms, which is why you have this book. Here we show you what to look for and the best way to improve your health. Let us look at what we stand to gain by choosing the mushrooms in this category.

- Improve brain health.

- Get help to fight inflammation.

- Support for the immune system.

- Boost the nervous system.

- Manage blood sugar.

- Useful antioxidants.

- Increase stamina and energy.

Mushroom to Improve Brain Health

Lion's Mane

Species name: Hericium erinaceus Health Benefit: Improves Brain Function

Important substances in Lion's Mane include beta-glucans, hericenones, and erinacines. Beta-glucans help lower cholesterol, boost the immune system, and improve blood sugar management. It protects against oxidative stress so that we remain protected against conditions like Parkinson's and Alzheimer's disease. The other substances hericenones and erinacines help improve the Nerve Growth Factor (NGF).

This species of mushrooms have a beautiful appearance with rows of long cascading shaggy spines resembling a lion's mane. It grows in the Northern Forest on the bark of hardwood trees.

Identification: Icicle-like teeth that hang from a central stalk. When it begins to grow, the teeth will be about one centimeter in length.

Cordyceps - Caterpillar Fungus

Species name: Cordyceps Militaris Health Benefit: Increases Energy and Lung Capacity

This mushroom is very important because it helps improve your energy. It increases ATP production by the action of precursors like cordycepin and adenosine. The ATP cells are the powerhouse of our bodies and athletes prefer this mushroom. In Chinese medicine, Cordyceps helps improve lung capacity and deal with seasonal allergies.

There are hundreds of different species in this Cordyceps family. This parasitic fungus preys on insects. By tradition, we see how people who use this fungus show an increase in endurance, stamina, appetite, and energy. The most popular one is Cordyceps sinensis that sells for over $20,000 per kilogram in Asia. We can get this rare combination of the caterpillars and fungi only on the Himalayan mountain range.

Even when the Chinese were not cultivating this mushroom, the traders were advertising it on the American market. This has increased the demand for this mushroom. So, what they sell there in America is probably those made with myceliated grain. Due to the high amount of grain and low amounts of beta-glucans, the potency of the mushroom is low.

Turkey Tail: Multicolored Mushroom

Species name: Trametes Versicolor Health Benefit: Boost for the Immune System

This mushroom stimulates cytokine production in the body. This boosts the natural killer cells and improves related immune functions.

Chapter Five

Psychedelic Mushrooms

Mushrooms have always remained surrounded in mystery since their discovery in the olden times. Magic mushrooms contain a naturally occurring hallucinogenic and psychoactive compound psilocybin and psilocin. These psychedelic substances are a Schedule I drug, meaning that the potential for misuse is high. Also, there is no accepted medical use in the United States. Dr. Albert Hofman first isolated this substance in 1958. People know him as the discoverer of lysergic acid diethylamide (LSD). The other names for magic mushrooms are philosopher's stones, liberty caps, golden tops, blue meanies, mushies, shrooms, agaric, Amani, and liberties.

Many poisonous mushrooms like the Horse Mushroom, Agaricus arvensis, or the field mushroom, Agaricus campestris, resemble the magic mushroom. A small mistake will result in a person falling sick or even dying.

Magic Mushroom

Liberty cap or magic mushroom is one that produces baeocystin and psilocybin, two of the most potent psychoactive substances. Hence, Psilocybe semilanceata, to give it the botanical name, is much sought after and widely distributed. Psilocybe means smooth head, and lanceata means spear-shaped. It belongs to the Strophariaceae family.

Description

Cap: This ranges from 0.4 to 2.2 cm in diameter and is cream-colored. There are striations on them that become bigger with age. You can see a distinct pimple on top of the cap.

Gills: They have free gills olive-gray. As they mature, they turn purple-black.

Stem: It stands 4-12 cm tall and is 2-3 mm thick.

Spores: These are smooth and ellipsoidal, 11.3-14.7 x 7-9 μm. The spore print is a very dark purple-brown.

Season: They grow throughout summer and autumn.

Odor and Taste: It has a musty odor. The taste is best avoided because Liberty cap is hallucinogenic.

Distribution

Magic mushroom has a widespread distribution. We see it grow in Finland, Estonia, Denmark, Belgium, Austria, the Netherlands, Switzerland, Spain, and Turkey in Europe. Also, it grows in Pakistan, Ukraine, and the United Kingdom. It also grows in the United States and Canada.

Toxicity

The mushroom has 1between 0.2% and 2.37% psilocybin, the psychoactive substance. The danger of misidentification exists. Picking a toxic mushroom will lead to dangerous results.

Sheathed Woodtuft

Kuehneromyces mutabilis is the botanical name of this fungus and other common names include Two-toned Pholiota and Brown Stew Fungus. It belongs to the Strophariaceae family.

Description

Cap: It is 3-7.3 cm in diameter, beginning convex and then becoming flattened with a broad umbo. Color is a bright tan that becomes pale ochre spreading from the middle giving the fungus a two-toned appearance. The flesh in the cap is thin and pale tan. Since this is a hygrophanous species, it dries out from the center, making the outer edge dark. This helps separate it from the poisonous Galerina marginata, which dries out from the edges to the center.

Gills: They remain crowded and adnate. In young mushrooms, they are pale ochre. This becomes cinnamon as it matures.

Stem: It has a ragged stem ring. The stem is smooth and pale above this. Below, it is fibrous, dark tan, and scaly. It becomes almost black at the base. Thickness is 5-10 mm, and it has a height of 3-8 cm. It is usually curved. The solid stem has pale tan flesh at the apex, which becomes dark brown at the base.

Cheilocystidia: It remains scattered and not abundant. Most of them are subcylindrical or narrowly lageniform. A few are subcapitate. The length is 20-40 μm width is 2.5 -7 μm.

Spores: These are smooth and, in a broad way, ellipsoidal. Size is 5.4 x 7.4 x 4.5 μm. It has an apical germ pore. The spore print is dark cinnamon brown to reddish ochre.

Odor/Taste: They are not distinctive.

Season: It grows throughout the year. It is plentiful in autumn and summer.

Distribution

Kuehneromyces mutabilis grows in Japan, Siberia, the Caucuses, Europe, North America, and Australia.

Toxicity

This mushroom is edible, but you must discard the tough stems. You can confuse it with Funeral Bell (Galerina marginata), so it is better to avoid taking a chance with this.

Chapter Six

Useful and Common Mushroom Types

M ost don't even know the difference between the mushrooms they see at the grocery store. The mushrooms you buy might have a nutty or sweet taste while some taste like lobster. You have seen the ones used for their medicinal value above that lower blood cholesterol, treat cancer, and other serious diseases. Now, we see the popular mushroom varieties you can use and be sure to cook them well.

Button Mushrooms

Botanical Name: *Agaricus bisporus*

This mushroom also has the names of white mushrooms or baby mushrooms. You are sure to find these because they are the most common variety of mushrooms available. They are small to medium in size. The size of their caps ranges between 5-7 cm. Sort truncated caps remain attached to the caps. The spongy and white caps are firm and rounded. When you bruise them, the white flesh will change color to pink and then brown.

On the underside of the white caps, there are small light brown gills. They remain under a white veil. We get dark brown spores from these gills and the stems are also edible, thick, smooth, and dense. Before cooking, they have a mild but crisp texture. When cooked, they display their typical earthy flavor with a chewy and tender texture.

You can get the white mushrooms all-around the year. The botanical classification for this is Agaricus bisporus. It is one of the most cultivated varieties and well used in this family. Sometimes people refer to it as Table Mushrooms, Cultivated Mushrooms, Champignon de Paris, and Common Mushrooms. They are recognizable by their color and age, but otherwise, they closely resemble the larger portobello mushroom or, the smaller cremini mushroom. Age-wise, button mushrooms are the youngest after which come the cremini mushrooms, and then we have the portobello mushrooms. White button mushrooms grow in grasslands and fields in many places world over. People also cultivate them, and the varieties in this group account for 90% of all cultivated mushroom varieties. Home cooks and chefs prefer them for their versatility and a mild taste.

Nutritional Value

Along with antioxidants, white button mushrooms contain selenium, potassium, manganese, folate, zinc, phosphorus, riboflavin, vitamin D, and amino acids.

Applications

We use white mushrooms, both cooked and raw while stewing, grilling, sautéing, roasting, and baking. In many instances, we use them instead of cremini mushrooms when the latter is not available. Sliced raw, the button mushrooms get combined with grain and green salad.

Cooked, we stuff them with crabs and grill them on skewers along with cheese and meat. They serve as a good appetizer or a baked tart. We add them to sauces, stews, soups, and stir-fries. They also combine well with artichokes. We can bake them into mushroom bread or chop them into ceviche.

They pair well with carrots, tomatoes, celery, basil, parsley, sage, lime, kimchi, fennel, ginger, onion, shallots, jalapenos, and potatoes. We use them along with meats like pork, beef, egg, or poultry. They combine with cream sauce, marinara sauce, balsamic vinegar, white wine, soy sauce, pecorino Romano, parmesan, mozzarella cheese, orzo, and rice.

It is possible to store them for up to a week in the fridge. Cover them up with moist paper towels to prolong their life.

Cultural Beliefs and Ethnic Values

People have used this variety of mushrooms since ancient times. So, they have a variety of uses and symbolism in different cultures. In Egypt, people believed that eating mushrooms will give them the secret power to eternal life. They cultivated the white mushrooms beneath the catacombs in Paris (hence the name Champignon de Paris). The Chinese used them to regulate the energy in the body and promote well-being.

Cremini Mushrooms

This variety of mushrooms (also called crimini mushrooms) belong to the same variety as button mushrooms. But they have a deeper flavor and a little brown color. All varieties of mushrooms were brown until 1926 when a Pennsylvanian farmer discovered a batch of white mushrooms. He began to sell them as a separate variety from then on.

The button and portobello mushrooms also belong to the cremini mushroom group. Where they differ is in how long they mature. White mushrooms are the youngest, and people grow them for their white color and soft texture. The cremini mushrooms that we are discussing here come between white and portobello mushrooms. We allow them to ripen a little more than the button mushrooms. This gives them a little stronger taste, but they remain like white mushrooms. Some people refer to them as baby portobello mushrooms.

When they age past the first two stages, they become portobello mushrooms. That is, they are mature mushrooms. They have a larger size, and the gills underneath the caps are denser. Cremini is the preferred variety for many people because they are "right" for their taste buds.

Identify this Variety

They have a little brownish appearance, but the stems are whitish. These mushrooms are more textured and shaggier than the white mushrooms. The gills are completely sheathed, and if you cut them across, they will be completely white with the beginnings of the gills visible.

It isn't advisable to forage for these mushrooms on your own. You will get Agaricus bisporus and their cousin Agaricus campestris, which is the field mushroom. To those who are not familiar with mushrooms, the Agaricus campestris will look the same as amanita specimens.

Aminta poisoning will not manifest up to five hours after eating them. It might even take a full day for the poisoning to show symptoms. If one delays it, the poison will have done considerable damage to the liver and kidneys. As little as half a cap can prove fatal. Since this variety of mushrooms are available in plenty in the malls, you don't need to try foraging and risk getting poisoned.

Growing the Mushroom

Since mushrooms don't require light (they have no chlorophyll), it is easy to grow them. Take compost to begin. Use any medium - straw, dry poultry waste, canola meal, water, or gypsum - and pasteurize the compost so that all fungal spores and bacteria already present get destroyed. Take concentrated mycelium culture and colonize it for several weeks. Mix this cremini spawn evenly in the compost.

Wait for a few days until the mycelium colonizes the compost. Add a layer of peat moss to give more moisture to the growing mushrooms. It will take 2-3 weeks for the mushroom to appear. You can also introduce mushrooms to shorten the waiting time. Once they appear, they grow very fast; they double in size once a day! In about four days, they reach the cremini stage and are ready for picking.

Picking the cremini mushrooms is a hand-operation. Cut cleanly at the stem with a knife. Avoid touching it many times, as this will increase the chances of bruising them. It is advisable to wait until it is time to cook them and harvest them. Cremini mushrooms are great because of their strong, earthy taste. For making soups, sauté them a little as they will taste better this way compared to raw.

Be sure to wash them before cooking since there might be a little dirt on them even if they come pre-washed. If you are not cooking them immediately, store them in the refrigerator immediately after you buy them. It will keep for about a week, so make sure you cook them before then. Freezing is not recommended, as this will alter its texture and taste. A good option is to sauté the mushrooms and cool them. Then, place them in airtight bags and put them in the freezer.

Also, do not pile up other foodstuffs on top of the mushroom. If you bruise them, they will spoil. Don't keep them next to food with strong odors like fish. The mushroom absorbs flavors, and so they will taste different. And, don't put them in the vegetable tray. There is far too much moisture there for the mushrooms to survive.

About whether you should eat raw mushrooms, many serve them along with vegetables as a salad.

Portobello Mushrooms

This is the grown version of the button mushroom. They are much larger than the white and cremini mushrooms. They have a meatier texture, but the flavor is still mild. Their caps are open, and you can see the dark gills beneath. Due to the big size, people use portobellos to make burgers. They stuff them with ingredients and bake them instead of frying them.

The portobello (also portabella) mushrooms are popular and delicious. People thought they were a separate species until recent research showed that they were only mature creminis.

Caution: Never eat portobellos raw; they contain hydrazine and agaritine. These are toxic substances.

Portabellas are the most common mushrooms used in burgers and pizzas. They can grow in any environment and all year-round. People didn't know about them until 1980 when they changed their name. This is the largest consumed mushroom in the world.

Health Benefits

Anti-cancer properties - There are many ingredients in portobellos such as grifolin, beta-glucans, lectins, and lentinan that inhibit the growth of cancer cells. More specifically, CLA a phytochemical inhibits cell proliferation. It induces apoptosis (make the cells cause suicide) in cancer cells and help in lipid metabolism. Portobello mushrooms cause a reduction in the size of tumors, according to a study conducted in mice. Another study showed how the beta-glucans were responsible for the death of cancer cells.

Good for the Blood - portobello mushrooms contain large amounts of copper and selenium. The human body uses copper for forming hemoglobin, and red blood cells useful for our respiration. It also helps in tissue repair, and improves our metabolism. This helps prevent fatigue and produces energy by breaking down oxygen. Selenium helps enhance the thyroid function that, in turn, helps us avoid hyperthyroidism. It also helps us overcome anxiety and depression by improving hormone activity.

Useful anti-inflammatory property - The antioxidants help control inflammation. It has fibers and L-ergothioneine that help fight inflammations in the body.

Cooking mushrooms will make them safe and more palatable though they will shrink in volume. Portobellos are good for roasting. When you cook them, the agaritine in them, being heat unstable, will disintegrate.

Oyster Mushroom

One of the favorites of mushroom lovers, the Pleurotus ostreatus mushroom, grows in tropical and temperate forests on dead and decaying wooden logs in shelf-like clusters. It has the name dhingri (Hindi), pearl oyster mushroom, and tree oyster mushroom. It belongs to the Pleurotus family and is a basidiomycete. It has an aroma of bitter almonds.

Parts of the Oyster Mushroom

There are three significant parts to the oyster mushroom.

    a.   Fleshy shell - a cap shaped like a spatula (pileus).

    b.   Stalk - short or long, central, or lateral (stipe).

    c.   Gills - underneath the pileus (lamellae).

This mushroom has a relatively large size, and gills are whitish. The stem is almost absent. In North America, it begins to sprout in October and goes on until early April. The kind of wood it grows on and the season it grows help separate one species from the next.

To Preserve the Environment

They kill bacteria and nematodes to such a large extent that conservationists use these mushrooms to clean environmental wastes. But the extent of the effort is not enough to clean the planet (through mycoremediation) or clean water (through mycofiltration).

Description of the Parts

Caps: The caps are 3-15 cm across with a broad, convex shape. They become flat to the top with a fan-shaped or kidney-shaped outline. They are bald with a greasy feel when they are wet or fresh. This is a sign of edible mushrooms. Color is pale to dark brown and fading slowly, and margins are a little enrolled in young mushrooms.

Stem: Lateral and rudimentary and almost absent when growing on trees, the stems are present when growing on logs. They grow up to 7 cm in length and 3 cm across with a tough, hairy, and velvety texture.

Gills: Run down to the stem, short and whitish to gray that become yellow with time. They have brownish edges and have black beetles growing inside.

The odor is distinct but hard to classify, and the flesh is white and thick. It doesn't change when we slice it. Spores are cylindric, ellipsoid, and 7-11 μm x 2-4 μm.

Habitat

This grows freely in the temperate and tropical regions of the world. But it is absent in the Pacific Northwest region where some other species grow. It grows all-around the year in the UK.

Cooking and Dishes

Korean, Chinese, and Japanese cuisine uses the oyster mushroom on its own or with other vegetables and ingredients to make soups and other culinary delicacies. It is good to eat on its own, as a soup, or as a stuffed dish. One must use the young mushroom when it is soft because they become as tough as they age. The taste is mild, and its odor resembles that of anise. While cooking, the mushrooms get torn up instead of getting sliced because this gives a better flavor.

Oyster dishes are popular in Kerala, the coastal town of India. They cultivate it in clear polythene bags that they layer with buns of hay. Now, they sow the spawn between the layers. Czech and Slovak's cuisine also sees the use of Oyster mushrooms where they eat it with or instead of meat.

Pearl oyster is useful to those who make mycelium furniture and mycelium bricks.

Warning: The sugar alcohol Arabitol present in Oyster might cause gastrointestinal disturbances in sensitive people.

Conclusion

Having come to the end of this wonderful collection of facts about "Poisonous Plants and Mushrooms," we are sure you will have learned new things. You can identify plants and mushrooms that will harm you. This will help you remain safe when you go foraging in the forest.

Keep the book handy in case you want to make some last-minute references. Hope you have a wonderful time discovering new plants and mushrooms.

Here is Wishing You All the Best on Your Journey of Discovery!

References

Daniels, E. (2019, December 18). 199 Poisonous Plants to Keep Away from Humans, Dogs & Cats. Retrieved August 01, 2020, from https://www.proflowers.com/blog/poisonous-plants

Mushroom poisoning. (2020, July 31). Retrieved August 01, 2020, from https://en.wikipedia.org/wiki/Mushroom_poisoning

Ogden Publications, I. (n.d.). A Beginners Guide to Foraging for Wild Mushrooms. Retrieved August 01, 2020, from https://www.motherearthnews.com/nature-and-environment/wild-mushrooms-zmaz87mazgoe

Poisonous Mushrooms: Some Facts, Myths, and Identification Information. (n.d.). Retrieved August 01, 2020, from https://www.mushroom-appreciation.com/poisonous-mushrooms.html

Poisonous plants. (n.d.). AccessScience. doi:10.1036/1097-8542.531500

Lightning Source UK Ltd.
Milton Keynes UK
UKHW021836271021
392931UK00009B/389

9 781955 786089